Airline Green Operations Strategies:

Emerging Research and Opportunities

Yazan Khalid Abed-Allah Migdadi
Qatar University, Qatar

A volume in the Advances in Mechatronics and Mechanical Engineering (AMME) Book Series

IGI Global
PUBLISHER of TIMELY KNOWLEDGE

Published in the United States of America by
IGI Global
Engineering Science Reference (an imprint of IGI Global)
701 E. Chocolate Avenue
Hershey PA, USA 17033
Tel: 717-533-8845
Fax: 717-533-8661
E-mail: cust@igi-global.com
Web site: http://www.igi-global.com

Library of Congress Cataloging-in-Publication Data

Names: Migdadi, Yazan Khalid Abed-Allah, 1979- author.
Title: Airline green operations strategies : emerging research and opportunities / by Yazan Khalid Abed-Allah Migdadi.
Description: Hershey, PA : Engineering Science Reference, [2020] | Includes bibliographical references and index. | Summary: "This book focuses on the green operational actions taken by airlines to protect the ecological environment from their activities"-- Provided by publisher.
Identifiers: LCCN 2019059600 (print) | LCCN 2019059601 (ebook) | ISBN 9781799842552 (hardcover) | ISBN 9781799851646 (paperback) | ISBN 9781799842569 (ebook)
Subjects: LCSH: Airlines--Management--Environmental aspects. | Aeronautics, Commercial--Management--Environmental aspects. | Ecological disturbances--Aviation--Management. | Greenhouse gases--Environmental aspects.
Classification: LCC HE9780 .M53 2020 (print) | LCC HE9780 (ebook) | DDC 387.7068/4--dc23
LC record available at https://lccn.loc.gov/2019059600
LC ebook record available at https://lccn.loc.gov/2019059601

This book is published in the IGI Global book series Advances in Mechatronics and Mechanical Engineering (AMME) (ISSN: 2328-8205; eISSN: 2328-823X)

British Cataloguing in Publication Data
A Cataloguing in Publication record for this book is available from the British Library.

Advances in Mechatronics and Mechanical Engineering (AMME) Book Series

ISSN:2328-8205
EISSN:2328-823X

Editor-in-Chief: J. Paulo Davim, University of Aveiro, Portugal

MISSION

With its aid in the creation of smartphones, cars, medical imaging devices, and manufacturing tools, the mechatronics engineering field is in high demand. Mechatronics aims to combine the principles of mechanical, computer, and electrical engineering together to bridge the gap of communication between the different disciplines.

The **Advances in Mechatronics and Mechanical Engineering (AMME) Book Series** provides innovative research and practical developments in the field of mechatronics and mechanical engineering. This series covers a wide variety of application areas in electrical engineering, mechanical engineering, computer and software engineering; essential for academics, practitioners, researchers, and industry leaders.

COVERAGE

- Nanomaterials and nanomanufacturing
- Medical robotics
- Control Systems Modelling and Analysis
- Control Methodologies
- Intelligent Sensing
- Computer-Based Manufacturing
- Mechanisms and machines
- Computational Mechanics
- Micro and nanomechanics
- Bioengineering Materials

IGI Global is currently accepting manuscripts for publication within this series. To submit a proposal for a volume in this series, please contact our Acquisition Editors at Acquisitions@igi-global.com or visit: http://www.igi-global.com/publish/.

The Advances in Mechatronics and Mechanical Engineering (AMME) Book Series (ISSN 2328-8205) is published by IGI Global, 701 E. Chocolate Avenue, Hershey, PA 17033-1240, USA, www.igi-global.com. This series is composed of titles available for purchase individually; each title is edited to be contextually exclusive from any other title within the series. For pricing and ordering information please visit http://www.igi-global.com/book-series/advances-mechatronics-mechanical-engineering/73808. Postmaster: Send all address changes to above address.

Titles in this Series

For a list of additional titles in this series, please visit:
http://www.igi-global.com/book-series/advances-mechatronics-mechanical-engineering/73808

Recent Technologies for Enhancing Performance and Reducing Emissions in Diesel Engines
J. Sadhik Basha (International Maritime College Oman, Sohar, Oman) and R.B. Anand (National Institute of Technology, Tiruchirappall, India)
Engineering Science Reference • © 2020 • 298pp • H/C (ISBN: 9781799825395) • US $215.00

Practical Approach to Substrate Integrated Waveguide (SIW) Diplexer Emerging Research and Opportunities
Augustine Onyenwe Nwajana (University of East London, UK) and Kenneth Siok Kiam Yeo (Universiti Teknologi Brunei Brunei)
Engineering Science Reference • © 2020 • 171pp • H/C (ISBN: 9781799820840) • US $195.00

Diverse Applications of Organic-Inorganic Nanocomposites Emerging Research and Opportunities
Gabriele Clarizia (Institute on Membrane Technology, National Research Council, Italy) and Paola Bernardo (Institute on Membrane Technology, National Research Council, Italy)
Engineering Science Reference • © 2020 • 237pp • H/C (ISBN: 9781799815303) • US $195.00

Handbook of Research on Artificial Intelligence Applications in the Aviation and Aerospace Industries
Tetiana Shmelova (National Aviation University, Ukraine) Yuliya Sikirda (Flight Academy of National Aviation University, Ukraine) and Arnold Sterenharz (EXOLAUNCH GmbH, Germany)
Engineering Science Reference • © 2020 • 517pp • H/C (ISBN: 9781799814153) • US $295.00

For an entire list of titles in this series, please visit:
http://www.igi-global.com/book-series/advances-mechatronics-mechanical-engineering/73808

IGI Global
PUBLISHER of TIMELY KNOWLEDGE

701 East Chocolate Avenue, Hershey, PA 17033, USA
Tel: 717-533-8845 x100 • Fax: 717-533-8661
E-Mail: cust@igi-global.com • www.igi-global.com

Table of Contents

Preface vii

Acknowledgment xiv

Chapter 1
Introduction to Airline Green Operations Strategy 1

Chapter 2
Airline Effective Direct Greenhouse Gases Emissions Strategies 17

Chapter 3
Airline Effective Indirect Greenhouse Gases Emissions Strategies 60

Chapter 4
Airline Effective Waste Management Strategies 87

Chapter 5
Airline Effective Water Consumption Strategies 110

Chapter 6
Airline Effective Noise Management Strategies 131

Chapter 7
Airline Effective Green Operations Strategy Patterns: Case Level Analysis 153

Chapter 8
Airline Effective Green Operations Strategy Patterns: Regional Level Analysis 172

Chapter 9
Airline Green Operations Strategy: Future Recommendations 190

Related Readings.. 213

About the Author 229

Index.. 230

Preface

Green operations strategy is one of the important topics under the umbrella of sustainable business practices. Sustainable business practices seek to do the present activities without harming ecological environmental, draining economic resources, or harming the society. So the current ecological, economics and social resources will be maintained for the future generations. The concept of operations strategy is a well-developed concept by the operations management literature. The early research contribution of Skinner (1969, 1972) tailored this concept as manufacturing strategy, then, this concept has been adopted by the next scholars as Lowson (2002) to be known as operations strategy.

According to operations strategy literature, the operations strategy content includes two main components: the operational actions and the operational competitive priorities. Accordingly, the operational units should take operational actions that will achieve the operations competitive priorities. The operational competitive priorities are the operational strategic objectives that related to products (goods or service) attributes. The scope of operational actions has been extended to cover the overall supply chain practices. Accordingly, the operational strategic actions includes all actions over the supply chain, as the supplier relationships management, the order fulfillment, the logistics management…etc.

The components of green operations content, which is the same as the components of operations strategy content. It includes both the green operational actions and the green indicators. The adopting of green operations strategy accordingly to "Triple Bottom Line" model will improve the operational performance as reducing cost, speeding-up the operational process, having more flexible operations, and improving the quality. In addition, the financial and the non-financial performance of the firms will be improved (Eklington, 1998). Accordingly, it is very important to adopt the most effective practices or the best practices of green operations strategy.

The green operational practices are affected by the green supply chain management (GSCM) literature, the concept of GSCM dates back to mid-1990s, when the initial literature of GSCM was emerged. Airline green operations strategy is a part of the green operations strategy field in general, and the aviation green operations strategy in particular. The aviation green operations strategy covers the practices of airlines, airports, aviation authority and any stakeholder as the local community and the municipalities. However, this book focuses on the green operational actions that taken by the airlines to protect the ecological environment from their activities. The airlines should be aware for the impact of its operational actions on the environment, as the density of direct, indirect GHG (Greenhouse Gases) emissions, the consumed water, the consumed energy, the consumed fuel, the generated wastes, and the severity of noise.

The concern about the environmental impact has raised-up over the last decades because of the severity of global warming phenomenon. The world leaders have met many times since mid-1990s at the United Nations Climate Change Conferences, to assess the progress in dealing with the climate change, negotiate the protocols as Kyoto and Paris, and establish the legal binding obligations for developed countries to reduce their greenhouse gasses emissions. Moreover, the United Nations General Assembly set 17 Sustainable Development Goals (SDG) in 2015, that intended to be achieved by the 2030, the main purpose of these goals is to achieve a better and more sustainable future for all. Many of these goals are related to protecting the ecological environment, as clean water and sanitation, affordable and clean energy, and climate actions. So, governmental, non-governmental organizations and business organizations are more enforced toward protecting ecological environment than before, since there is a global ecological environment framework.

Aviation sector is one of the sectors that affected by these global ecological environment framework. The environmental impact of airlines has increased over the last 25 years because of the increase in demand (Lee et al., 2009). CO2 emissions of airlines are expected to increase by further 45% between 2014 and 2035. The projected Australian civil aviation sector CO2 equivalent by 2020 is expected to be increased by 214% above 1990 level (BITRE, 2009). CO2 emissions of aviation sector in European region reached 163 million tonnes in 2017, by increasing of 16% in comparison with 2005. The forecasted CO2 emissions of 2040 is expected to reach 224 million tons by increasing of 59% in comparison with 2005 (EASA, 2019).

Over the last decades, many environmental targets were set for this sector by the professional institutions as IATA and ICAO. In addition, these institutions, the aviation authorities, the airports and the airlines have launched many initiatives. ICAO in 2004 adopted three major environmental goals, as reducing the number of people affected by the significant aircraft noise, reducing the impact of aviation emissions on local air quality, reducing the impact of greenhouse gases emissions on the global climate (ICAO, 2019). The future green targets of aviation as adopted by IATA is improving the average efficiency in fuel by 1.5% per year from 2009 to 2020, and reducing the net aviation CO2 emissions of 50% by 2050 relative to 2005 level. (IATA, 2019).

Accordingly, the aviation industry is responsible about making more actions to reduce the negative environmental impact. All stakeholders must make a genuine commitment to transform the transport system (UN-Secretary-General-High-Level Advisory Group on Sustainable Transportation, 2015). IATA has setup several programs to assist airlines to improve their environmental performance as the alternative fuel, IATA carbon offset program, the environmental assessment, and FRED+ (IATA, 2019). ICAO developed a range of policies, standards and guidance material for the application of integrated measures to address the noise and emissions, embracing technology improvements, operating procedures, proper organization of air traffic, and appropriate airport and land-use planning (ICAO, 2019).

This is a timely book, which helps the airlines to adopt the most effective green operational strategies. In addition, other practitioners who concern about tracking the environmental impact of airlines and developing policies, regulations or directing airlines toward the most effective practices will benefit from this book. These practitioners could be the airlines public management authorities, International Civil Aviation Organization (ICAO), International Air Transport Association (IATA), Civil Air Navigation Services Organization (CANSO), Air Transport Association, National Aeronautic Association (NAA), National Traffic Controllers, Flight Safety Foundation (FSF), environmental societies, aircraft producers, airlines environmental quality systems developer, and consultants. Moreover, the academics as airlines management professors, lectures and researchers have in-depth analysis for the airlines effective green practices, which could be utilized for teaching purposes, or for developing models and hypotheses for the future studies of effective green practices.

This book adopted a quantitative research methodology. The data of 23 airlines from five regions was collected. The facts about the green performance related to five green indicators and actions were collected. The facts were related to the direct GHG emissions, the indirect GHG emissions, the waste management, the water consumption, and the noise management. The data was collected over the period 2013 to 2016, and collected from the published sustainability reports of airlines. These reports were retrieved from Global Reporting Initiatives (GRI) database.

This book is issued in nine chapters. The first chapter discussed the airlines environmental initiatives and challenges. Therefore, a general introduction of green operations strategy configuration and content was presented, and the research gaps and contributions of this book were discussed. The second chapter aimed to report the effective direct GHG emissions strategies. To achieve this aim, the flight and ground support practices of 22 airlines from five regions were investigated. This chapter reported eight effective flight operations strategic patterns and two effective ground operations strategic patterns.

The third chapter aimed to report the effective indirect GHG2 emissions strategies. To achieve this aim the practices of 18 airlines from five regions were investigated. The chapter reported eight effective strategic patterns. The fourth chapter aimed to report the effective wastes management strategies. To achieve this aim, the waste management practices of 14 airlines from four regions were investigated, this chapter reported four onboard and ground strategy patterns, and three hybrid patterns.

The fifth chapter aimed to report the effective water consumption strategies. To achieve this aim, the water saving practices of 14 airlines from four regions were investigated, this chapter reported five effective strategic patterns. The sixth chapter aimed to report the effective noise management strategies of airlines. To achieve this aim, the practices of 18 airlines from five regions were investigated. This chapter reported seventeen effective strategic patterns.

The seventh chapter aimed to report the effective green operations strategy patterns. To achieve this aim, the green operations strategy patterns of 23 airlines from five regions were investigated. This chapter depended on the results of previous chapters. Accordingly, the reported effective green operations strategy under each indicator were used to report the effective green operations strategy patterns. The special tailored data analysis techniques were used for this purpose. The chapter reported eight effective strategy patterns.

The eighth chapter aimed to report the airline effective green operations strategy patterns that adopted by each region. To achieve this aim, green practices of 23 airlines from five regions were investigated, the data used in this chapter was the effective green operations strategy that adopted by each airline, which is the result of the seventh chapter. This chapter reported the effective green strategy patterns that adopted by each region.

The last chapter aimed to recommend future improvement on airline green operations strategy patterns to have more effective strategy patterns. To achieve this aim, the ineffective green operations strategy of 12 airlines from five regions were analyzed, and the changes on the ineffective strategies were recommended. The average overall index of all green indicators will be improved to be positive. All indexes of all regions will be improved to be positive, the number of effective patterns after improvement will be seven. In addition, the green index achieved after improvement will be moderate.

Airline effective green operations strategy of each regions have been reported by very limited studies (e.g. Migdadi, 2018, 2019, 2020). Some of these studies have reported the airline membership for a particular green strategy pattern, with limited concern about investigating in-depth the strategy patterns that adopted by each region. The most of previous studies have reported the effective green operations strategies, but these studies have not concerned about recommending future changes in the current adopted strategies to have more effective strategy patterns (e.g. Migdadi, 2016; Migdadi & Elzaqaiba, 2018; Migdadi, 2018; Migdadi & Omari, 2019; Migdadi, 2019; Migdadi, 2020).

Moreover, the previous studies have investigated the green actions that will reduce GHG emissions. The most of studies have focused on reporting the flight operations actions. The limited studies that investigated indirect GHG2 emissions have reported the general strategy patterns that adopted by airlines, so reporting the effective practices of GHG2 emissions was a part of the general scope of the studies. The concern about reporting the effective water saving strategic patterns have not been the main research issue by the previous studies, and the strategic patterns of airlines noise management have not been investigated by the previous studies.

This book aimed to bridge the previous research gaps by reporting the effective green operations strategy in reducing the direct GHG emissions, the indirect GHG emissions, the wastes generated, the water consumption and noise. Moreover, reporting the effective strategy patterns at the cases and the regional level. Finally, recommending some changes in the current ineffective green operations strategy patterns to have more effective strategy patterns in future.

Yazan Khalid Abed-Allah Migdadi
Qatar University, Qatar

REFERENCES

BITRE. (2009). *Greenhouse gas emissions from Australian transport: projections to 2020*. Working paper 73. https://www.bitre.gov.au/publications/2009/files/wp_073.pdf

EASA. (2016). *European aviation environmental report 2016*. European Airline Safety Association. https://www.eurocontrol.int/sites/default/files/2019-06/eaer-2019_0.pdf

Eklington, J. (1998). *Cannibals with forks: The Triple Bottom Line of the 21st century*. Stoney Creek, CT: New Society Publishers.

IATA. (2019). *Improving environmental performance*. https://www.iata.org/whatwedo/environment/Pages/index.aspx

ICAO. (2019). *Environmental protection*. http://www.ICAO.int/environmental-protection/pages/default.aspx

Lee, D. S., Fahey, D. W., Forster, P. M., Newton, P. J., Wit, R. C., Lim, L. L., ... Sausen, R. (2009). Aviation and global climate change in the 21st century. *Atmospheric Environment*, *43*(22-23), 3520–3537. doi:10.1016/j.atmosenv.2009.04.024

Lowson, R. H. (2002). Strategic Operations Management—The New Competitive Advantage? *Journal of General Management*, *28*(1), 36–56. doi:10.1177/030630700202800103

Migdadi, Y. K. A. A. (2016). Identifying the best practices in green operations strategy of leading mobile phone producers. *International Journal of Business Excellence*, *9*(1), 92–112. doi:10.1504/IJBEX.2016.073377

Migdadi, Y. K. A. A. (2018). Identifying the best practices of airlines' green operations strategy: A cross-regional worldwide survey. *Environmental Quality Management*, *28*(1), 21–32. doi:10.1002/tqem.21575

Migdadi, Y. K. A. A. (2019). Identifying the effective taxonomies of airline green operations strategy. *Management of Environmental Quality*.

Migdadi, Y. K. A. A. (2020). Green Operational Strategy for Airlines: Content and Regional Analysis. In Cases on Green Energy and Sustainable Development (pp. 193-229). IGI Global.

Migdadi, Y. K. A. A., & Elzzqaibeh, D. A. S. I. (2018). The evaluation of green manufacturing strategies adopted by ISO 14001 certificate holders in Jordan. *International Journal of Productivity and Quality Management*, *23*(1), 90–109. doi:10.1504/IJPQM.2018.088610

Migdadi, Y. K. A. A., & Omari, A. A. (2019). Identifying the best practices in green operations strategy of hospitals. *Benchmarking: An International Journal*, *26*(4), 1106–1131. doi:10.1108/BIJ-09-2017-0242

Skinner, W. (1969). Manufacturing-missing link in corporate strategy. *Harvard Business Review*, *47*(3), 136–145.

Skinner, W. (1974). The focused factory. *Harvard Business Review*, *52*(3), 113–121.

UN-Secretary-General-High-Level Advisory Group on Sustainable Transportation. (2015). *Mobilizing sustainable transport for development*. https://sustainabledevelopment.un.org/content/documents/2375Mobilizing%20Sustainable%20Transport.pdf

Acknowledgment

First and foremost, I praise and thank the God, the Almighty, for his blessings throughout the work of my book, to complete the book successfully.

I would like to express my deep and sincere gratitude to IGI Global Publisher for the support provided to publish my book.

I am extending my thanks to all scholars whose intellectual contributions are cited on my book.

I would like to express my special thanks to the reviewers for their helpful comments in improving the book content.

My thanks go to my wife Bayan for her continuous encouragement and motivating me.

Yazan Khalid Abed-Allah Migdadi
Qatar University, Qatar

Chapter 1
Introduction to Airline Green Operations Strategy

ABSTRACT

This chapter introduces this book. To achieve this aim, this chapter discussed the environmental initiatives and challenges of airlines. Then, a general introduction of green operations strategy configuration and content was presented. Finally, the research gaps and contributions of the book were discussed. This book aimed to bridge the current research gaps by reporting the effective green operations strategy in reducing direct GHG emissions, indirect GHG emissions, waste, water consumption, and noise. Moreover, the book reports the effective strategy patterns at case and regional levels. Finally, the book recommends some changes in the current ineffective green operations strategy patterns to have more effective strategy patterns in the future.

AIRLINE ENVIRONMENTAL INITIATIVES AND CHALLENGES

The total economic impact of the aviation is about 3.5% of the world's GDP, about USD 2.7 trillion, which supports 62.7% million jobs worldwide (Gitten et al., 2017). Aviation of Latin America contributed more than USD 167 Billion and provides 5.2 million jobs (ECLAC, 2017). However, Asia aviation contributes about USD 218 billion revenue (Herdman, 2018). Middle East aviation contributes about USD 130 billion to GDP (Air Transport Aviation

DOI: 10.4018/978-1-7998-4255-2.ch001

Group, 2019). Asia pacific remains the largest region of activity with 34% of world traffic (Herdman, 2018), followed by Europe and North America with 27% of world traffic for each region. Middle East region represents 9% of the world traffic. Latin America region accounts for 5%, while the remaining 2% of the world traffic is under taken by Africa region. The air traffic volumes will be doubled by 2032, characterised by 4.6% annual growth rate for passenger traffic (Gittens et al., 2017).

The environmental impact of airlines has increased over the last 25 years as a result of increase in demand (Lee *et al.,* 2009). CO2 emissions of airlines are expected to increase by further 45% between 2014 and 2035. The projected Australian civil aviation sector CO2 equivalent by 2020 is expected to be increased by 214% above 1990 level (BITRE, 2009). CO2 emissions of aviation sector in European region reached 163 million tonnes in 2017, by increasing of 16% in comparison with 2005. The European region forecasted CO2 emissions of 2040 is expected to reach 224 million tons by an increase of 59% in comparison with 2005, however, there is a reduction in noise per aircraft by 14% in comparison with 2005 (EASA, 2019).

The energy conservation output and emission reduction of airports buildings are not sufficient and should be expanded in future (Jun & Jie, 2018). The aviation industry will be more responsible about making more actions to reduce the negative environmental impact; all stakeholders must make a genuine commitment to transform the transport system (UN-Secretary-General-High-Level Advisory Group on Sustainable Transportation, 2015). Despite the early concern about managing the cabin wastes, until now most airlines and catering companies have been recycled very little. In addition, wastes obtained is typically of low quantity due to the mix of multiple wastes fractions. There is some changes in the trend of wastes recycling over the last few years. The efforts are still single initiatives, lacking comprehensive and holistic concern (Blanca-Alcubilla et al., 2018).

Airlines took some actions which achieved some progressing in term of water saving. The achieved saving was on average -2.6%. The most effective cases achieved a saving by -8% (Migdadi, 2018), -3.75% and -14.16% (Migdadi, 2020). Despite the average level of noise is lower than before and lower than road and rail transportation noise, the frequency of aircrafts movements, and hence noise event has generally increased (Glasgow Airport, 2014). The noise tends to cover larger geographical areas and is more difficult to mitigate when compared to the road and rail noise (Civil Aviation Authority, 2014).

Different green initiatives have been taken by aviation sector in different regions. For example, Brazil PBN procedures have been implemented since

2012, noise reduction through stable descent trajectories and reduction of approximately 203,000 metric tonnes of jet fuel or 64, 000 of CO2 per year (Sameh & Scavuzzi, 2016). In the USA, the aviation industry implemented an action plan to reduce GHG emissions, which includes improving airframe and engine technology, operating more efficient aircraft, and developing and deploying of alternative jet fuels (ICAO, 2015). The number of Europe airports participating in the airport carbon accreditation program has increased from 92 to 133, and the number of airport reaching CO2 neutral status rose from 20 to 37 (EASA, 2019). African countries continue to grapple unsuccessfully with the idea of sustainable aviation. The major challenge facing the African region is that, the unavailability of technology that foster the sustainability. In addition, if the technologies are available, there will be a dearth of human capacity to maintain or optimally utilize them (Olawuy, 2016).

IATA has a professional committee advising the IATA on environmental policies and strategies. This committee is IATA's Environmental Committee (ENCOM). The recommended policies by this committee includes: climate change, aircraft noise, local air quality and illegal wild life traffic king (IATA, 2019a). The goals of the polices is to ensure the proportionate and coherent environmental policies are implemented by governments around the world to support sector's sustainability and allow environmental improvement to be achieved in a cost-effective way (IATA, 2019b). IATA has set up several programs to assist airlines to improve their environmental performance such as alternative fuel, IATA carbon offset program, environmental assessment and FRED+ (IATA, 2019a).

ICAO has a Council's Committee on Aviation Environmental Protection (CAEP). This committee consists of members and observers from states, intergovernmental and non-governmental organizations representing aviation industry. Also, maintains close relations with other UN policy-making bodies. ICAO developed a range of policies, standards, and guidance material for the application of integrated measures. To address noise and emissions, embracing technology improvements, operating procedures, proper organization of air traffic, and appropriate airport and land-use planning (ICAO, 2019).

ICAO in 2004 adopted three major environmental goals: reduce the number of people affected by significant aircraft noise, reduce the impact of aviation emissions on local air quality and reduce the impact of greenhouse gases emissions on the global climate (ICAO, 2019). The future green targets of aviation as adopted by IATA are improving the average efficiency in fuel by 1.5% per year from 2009 to 2020, and reducing the net aviation CO2 emissions of 50% by 2050 relative to 2005 level. (IATA, 2019a). Accordingly, the aviation industry is responsible about making more actions to reduce the negative environmental impact. All stakeholders must make a genuine commitment to transform the transport system (UN-Secretary-General-High-Level Advisory Group on Sustainable Transportation, 2015).

AIRLINE GREEN OPERATIONS STRATEGY CONFIGURATION AND CONTENT

According to "Triple Bottom Line" approach of sustainable management, the building blocks of sustainability are economic, social and environmental factors. There is mutual relationships between these factors (Eklington, 1998). Accordingly, the environmental factor is one of the building blocks of the sustainability. The theory of green operations is studied from two perspectives: the first is strategic perspective. According to this perspective, the operations are the source for the competitive advantage of the organization since it is responsible about making products or services. The second approach focuses more on the operational environmental decisions such as design for environment, green supply chain management, reverse logistics, and quality management (Gupta & Sharma, 1996). Regardless of the perspectives adopted, green operations strategy is a stream of actions related to organization operations, which affects the corporation environmental competitive advantage. The operational decisions related to many dimensions, which are discussed in the following section. However, the judgement regarding the environmental competitive advantage could be traced through comparing the environmental performance of a particular organization with the competing organization, so it is the relative environmental performance of a particular organization in comparison with other competing organizations. The performance will be assessed according to many indicators, which are discussed in the next section.

The organization could adopt many actions and achieve many green indicators. Therefore, the green strategy will be formed as a strategy pattern or configuration. Strategy configuration is a building block of attributes and actions. These configurations are called strategy patterns. Two approaches have been adopted to develop operations strategy configurations. The first approach is the typology approach and the second is the taxonomies approach. According to typology approach, the strategy configurations are ideal models or patterns that could be developed by investigating limited number of best practices, or it could be developed by using literature. The typologies are used as a reference for comparing the practices of cases in order to modify practices based on typology attributes. In addition, the typologies could be used as a classification scheme for the practices and the typologies could be used as initial or ground theories which will be used later for more extended quantitative research (Migdadi, 2011a, 2013).

The second strategy configuration approach is the taxonomies of strategy. It is a quantitative approach, based on a quantitative analysis of collected facts about the investigated cases. Accordingly, the taxonomies approach results present both the effective and ineffective strategy patterns. Therefore, it is more realistic in comparison with typologies approach (Migdadi, 2011a). The widely adopted quantitative analysis approach adopted by the previous studies was clustering analysis in general and K-means clustering analysis in particular (e.g. Migdadi, 2011b, Migdadi, 2012; Migdadi, 2014; Alkhatib & Migdadi, 2018; Migdadi & Elzaqaiba, 2018; Migdadi, 2018; Alkhatib & Migdadi, 2019; Migdadi, 2019; Migdadi & Omari, 2019).

AIRLINE GREEN PERFORMANCE INDICATORS DEFINITION AND MEASURES

The airline green indicators can be classified into direct GHG1 emissions, indirect GHG2&3 emissions, fuel consumption, energy consumption, wastes generated, water consumption and noise (Migdadi, 2018, 2019, 2020). The airlines greenhouse gas emissions include the emissions of carbon dioxide CO_2, the emissions of NOx, emissions of sulphate arising from Sulphur in fuel, water vapor and formation of cirrus clouds (Bofinger & Strand, 2013). These emissions can be classified according to its effect on the climate into direct emissions, indirect emissions, emissions of aerosols, formation of contrails and cirrus clouds. The direct GHG emissions according to the previous classification includes such emissions as CO and water vapor. However, the indirect GHG emissions includes NOx emissions, the emissions of aerosols, which have cooling or warming effect as sulfates and soot. The formation of contrails and cirrus clouds are emissions, which occur in the troposphere and lower stratosphere (Fuglestvedt et al., 2008).

The GHG emissions of airlines can be classified into direct and indirect emissions. The direct emissions or what is called GHG1 emissions are the metric tones of CO2 or equivalent CO2 emissions of aircrafts and ground support operations. However, the indirect emissions can be classified into

GHG2 and GHG3. The indirect GHG emissions are consequence of the particular company (airline) activities, but occur at sources owned or controlled by another company. GHG2 emissions, or what is called scope 2 emissions, are the emissions from the purchased and consumed heat, electricity or steam (Sotos, 2015). It is the metric tons of indirect CO2 emissions of electricity, power and heat from the direct billing of owned or leased facilities (Migdadi, 2018, 2019, 2020).

The airlines wastes can be classified into on-board (cabin) and ground wastes, industrial wastes and hazard wastes (Migdadi, 2018; 2019; 2020). The composite of cabin wastes includes clean paper (mainly newspaper) was the largest component. Plastic as drinking cups and food covers, which was the next major component. And glasses (Li et al., 2003; Hondett, 2010). Accordingly, the most significant waste is paper. The newspapers, magazines, and advertisements count almost 9.21% to18.91% of the paper waste. Disk paper count 0.96% to 2.65% of the paper waste. The containers of coffee and tea bags count 4.6% to 14.78%. And toilet paper, which counts 3.85% to 11.90% of the paper wastes (Mehta, 2015). Industrial wastes are mainly the aircrafts parts generated from retired, stored, abandoned aircrafts, and aircraft previously involved in accidents (IATA, 2018). Hazard wastes are substances that because of its quantity, physical, concentration or infectious characteristics. On the other hand, when it is improperly treated, stored, transported or disposed cause series illness, or pose a substantial or potential hazard to human health or for environment. In the aircraft industry, it can originate from the service of repair of aircraft components (Wieher, 2014).

Water consumption is an indicator which measured as cubic meter (m^3) of water consumed per passenger of an aircraft (Wyman, 2012), or an aircraft movement over a period of time (Baxter, Srisaeng & Wild, 2018a). Noise is unwanted level of sound (CANSO, 2015). The aviation noise is caused by air passing over the aircraft body, which causes turbulence and friction. The noise is caused by the sounds of engine moving parts and sound of air being expelled at high speed (Glasgow Airport, 2014). The aviation noise metrics can be classified into single event based metrics or the individual aircraft noise over a short period of time. Other metrics of noise are the time based noise metrics or the average noise over a defined period. Threshold based metrics, which is the total number of time a defined noise threshold is exceeded in a specific period as 24 hours (CANSO, 2015).

Different indexes have been developed for the green airline practices by different institutions such as Atmosfair and Dow Jones. Atmosfair is a non-profit organization for combating climate change. It was founded in 2004 for a research project of the German Federal Ministry for the Environment. Atmosfair developed annual ranking of airlines according to index called Atmosfair Airline Index. This index is developed by calculating of CO2 per net load Km. for each flight. Which is based on aircraft type, engine, seat, cargo capacity and load factor. Then, comparing with best-case flight. Next, determine of the city pair efficiency points of an airline, finally, ranking of airlines by the global efficiency points. The source of data is ICAO, IATA, OAG and FlightGlobal (Atmosfair, 2018).

Dow Jones Sustainability Index (DJSI) is a family of best-in-class benchmarks for investors who have recognised that sustainable business practices are critical to generate long-tern shareholders value, who wish to reflect their sustainability convictions in their investment portfolio. The family was launched in 1999 as the first global sustainability benchmark and tracks the stock performance of the world leading companies in term of economic, environmental and social criteria. S&P Dow Jones and Rebeco SAM created it. The data is collected by using a questionnaire survey. In 2018 about 3500 companies were invited to participate in this survey. The data was collected from different industries (real estate, aerospace, commercial services, food products, life science and services). In addition, the data was collected from different regions such as Europe, North America, and Asia (S&P Dow Jones Indices & RobecoSAM, 2019).

AIRLINE GREEN OPERATIONS ACTIONS DEFINITION AND CATEGORIES

The airline green actions can be classified into direct GHG1 emissions actions, indirect GHG2 & 3 emissions actions, wastes management actions, water management actions and noise management actions (Migdadi, 2018, 2019, 2020). The airline direct GHG emissions actions are those related to flight and ground supporting actions (Migdadi, 2018). These actions are classified into flight operations and ground operations. The flight operations actions are classified into different categories as technological advancements, operational improvements and alternative fuels (El-Takriti, Pavlenko & Searle, 2017). Another adopted classification scheme classified actions into operations, aircraft design, and alternative fuel (McCollum, Gloud & Greene, 2009). Other classifications combined them under two main categories as technology and process-based (Yan, Cui & Gil, 2016).

The operational actions of indirect GHG2 emissions can be classified into facilities and buildings energy saving, facilities and buildings accreditation, facilities and buildings sustainable energy, upgrading and replacing facilities (Migdadi, 2018, 2019 & 2020). The aviation wastes reduction actions involves normally waste minimization (waste avoidance or prevention, waste reuse, recycling, energy recovery), as well as waste disposal (London Stansted Airport, 2010; Reiss & Becker, 2018; Baxter, Srisaeng & Wild, 2018). The actions of airlines water saving can be classified into recycling and recovery of maintenance water, saving of maintenance water, recycling and recovery of building and facilities water, and saving of building and facilities water (Migdadi, 2918; 2020).

The widely adopted approach of noise management is "Balanced Approach to Aircraft Noise Management", which is recommended by ICAO. According to this approach, the noise of each airport will be measured and analyzed according to reduction of noise through tracking four measures dimensions: reduction of noise at source, land-use planning and management, noise abatement operations procedures and operations restrictions. These policies are published in ICAO Doc. 9829, "Guidance on the Balanced Approach to Aircraft Noise Management" (ICAO, 2019a).

RESEARCH GAPS AND CONTRIBUTION OF THIS BOOK

The most of previous studies that investigated the green actions that will reduce GHG emissions have focused on reporting the flight operations actions (e.g. Chapman, 2016; El-Takriti, Pavlenko & Searle, 2017; Lin, Zahou, Zahou & Wang, 2017; Lee, Tsai, Yang & Lin, 2018). The actions have been traced by most studies were limited in their scope. Most of the previous studies have reported the actions with limited concern about the different strategic patterns adopted. The most of previous studies have investigated the green actions that will reduce GHG2 emissions in the airports (e.g. Ortega Alba & Manana, 2016; Sukumaran & Suhakar, 2017; Baxter, Srisaeng & Wild, 2018b,c; Xu & Deng, 2018; Baxter, Srisaeng & Wild, 2019). The limited studies that investigated GHG2 emissions have reported the general strategy patterns adopted by airlines. So reporting the effective practices of GHG2 emissions was a part of the general scope of the studies (e.g. Migdadi, 2018, 2019 & 2020).

The most of previous studies have focused on reporting the on-board (in-flight) wastes reduction actions (e.g. Blanca-Alcubilla et al., 2018; Tofalli,

Loizia & Zorpas, 2018). The actions have been traced by most studies were limited in their scope. The most of previous studies have reported the actions with limited concern about the different strategic patterns adopted. The practices of airlines were investigated for airlines in one country or regions for some selected cases (e.g. Blanca-Alcubilla et al., 2018; Tofalli, Loizia & Zorpas, 2018). Also, the previous studies have investigated the water saving action of airports (e.g. Carvalho et al. 2013; Shamasky & Biochenko, 2016; Baxter, Srisaeng & Wild, 2019). The concern about reporting the effective water saving strategic patterns have not been the main research issue by the previous studies.

Many studies have investigated the noise. Some of them have investigated the abatement operations procedure (e.g. Fillipone, 2017; Gagliardi Teti & Licitra, 2018; Pretto, Giannattas, De Gennavo & Zanon, 2019). Some studies have investigated the reduction of noise at source (e.g. Torija et al., 2019) and other studies have investigated the land-use planning (e.g. Flores, Asensio, Gagliardig & Licitra, 2019). Very limited have investigated the airlines green operations strategies (e.g. Migdadi, 2018, 2019, 2020). These studies have investigated the general strategic patterns. The strategic patterns of airlines noise management have not been investigated by previous studies. Reporting the effective taxonomies of airlines operations strategy in general and green operations strategy in particular is still limited. For example, some of the previous studies have reported the effective capacity strategy (e.g. Alkhatib & Migdadi, 2018) and effective scheduling strategy (e.g. Alkhatib & Migdadi, 2019). The effective green operations strategy contribution have been reported by some studies as the study of Migdadi (2018, 2019).

Airline effective green operations strategy of each regions have been reported by very limited studies (e.g. Migdadi, 2018, 2019, 2020). Some of these studies have reported the airline membership for a particular green strategy pattern, with limited concern about investigating in-depth the strategy patterns that adopted by each region. The most of previous studies have reported the effective green operations strategies, but these studies have not concerned about recommending future changes in the current adopted strategies to have more effective strategy patterns (e.g. Migdadi, 2016; Migdadi & Elzaqaiba, 2018; Migdadi, 2018; Migdadi & Omari, 2019; Migdadi, 2019; Migdadi, 2020).

Accordingly, the aim of this book is to bridge the previous research gaps by achieving the following objectives:

1. Reporting the effective green operation strategies in reducing direct GHG emissions, indirect GHG Emissions, waste, water consumption and noise.
2. Reporting the effective green operations strategy patterns that adopted by airlines.
3. Reporting the effective green operations strategy patterns that adopted by each region's airlines.
4. Recommending the changes in ineffective green strategy patterns to have more effective strategy patterns in the future.

Chapter 2, chapter 3, chapter 4, chapter 5, and chapter 6 achieved the first objective. However, chapter 7 achieved the second objective. The third objective was achieved by chapter 8. Finally, chapter 9 achieved the last objective.

REFERENCES

Air Transport Aviation Group. (2019). *Middle East*. https://aviationbenefits.org/around-the-world/middle-east/

Alkhatib, S. F. S., & Migdadi, Y. K. A. A. (2018). Airline capacity strategies: Worldwide analysis, taxonomies and regional comparison. *International Journal of Advanced Operations Management*, *10*(4), 281–303. doi:10.1504/IJAOM.2018.097261

Alkhatib, S. F. S., & Migdadi, Y. K. A. A. (2019). Taxonomies of airline scheduling strategies: Cross-regional worldwide analysis. *International Journal of Services and Operations Management*, *33*(1), 26–48. doi:10.1504/IJSOM.2019.099653

Atmosfair. (2018). *Atmosfair airline index 2018*. Berlin: Germany. https://www.atmosfair.de/wp-content/uploads/aai2018-englischfarbe_final_mn.pdf

Baxter, G., Srisaeng, P., & Wild, G. (2018a). An Assessment of Sustainable Airport Water Management: The Case of Osaka's Kansai International Airport. *Infrastructures*, *3*(4), 54. doi:10.3390/infrastructures3040054

Baxter, G., Srisaeng, P., & Wild, G. (2018b). Sustainable airport energy management: The case of kansai international airport. *IJTTE. International Journal for Traffic and Transport Engineering*, *8*(3), 334–358. doi:10.7708/ijtte.2018.8(3).07

Baxter, G., Srisaeng, P., & Wild, G. (2018c). An assessment of airport sustainability, Part 2—Energy management at Copenhagen Airport. *Resources*, *7*(2), 1–27. doi:10.3390/resources7020032

Baxter, G., Srisaeng, P., & Wild, G. (2019). Environmentally sustainable airport energy management using solar power technology: The case of Adelaide Airport, Australia. *IJTTE. International Journal for Traffic and Transport Engineering*, *9*(1), 81–100. doi:10.7708/ijtte.2019.9(1).07

BITRE. (2009). *Greenhouse gas emissions from Australian transport: projections to 2020*. Working paper 73. https://www.bitre.gov.au/publications/2009/files/wp_073.pdf

Blanca-Alcubilla, G., Bala, A., Hermira, J. I., De-Castro, N., Chavarri, R., Perales, R., ... Fullana-i-Palmer, P. (2018). Tackling International Airline Catering Waste Management: Life Zero Cabin Waste Project. State of The Art And First Steps. *Detritus*, *3*, 159.

Bofinger, H., & Strand, J. (2013). *Calculating the carbon footprint from different classes of air travel*. The World Bank. doi:10.1596/1813-9450-6471

CANSO. (2015). *Managing the impact of aviation: a guide for airport operations and air navigation service providers*. https://www.canso.org/sites/default/files/Managing%20the%20Impacts%20of%20Aviation%20Noise_HQ.pdf

Carvalho, I. D. C., Calijuri, M. L., Assemany, P. P., Silva, M. D. F. M., Moreira Neto, R. F., Santiago, A. D. F., & Souza, M. H. B. D. (2013). *Sustainable airport environments: a review of water conservation practices in airports*. Academic Press.

Chapman, M. (2016). Sustaining Reductions in Aircraft Emissions for Canada's Major Airlines. In *Managing in a VUCA World* (pp. 175–193). Cham: Springer. doi:10.1007/978-3-319-16889-0_12

Civil Aviation Authority. (2014). *Managing aviation noise*. Kingsway. https://publicapps.caa.co.uk/docs/33/CAP%201165%20Managing%20Aviation%20Noise%202.pdf

EASA. (2016). *European aviation environmental report 2016*. European Airline Safety Association. https://www.eurocontrol.int/sites/default/files/2019-06/eaer-2019_0.pdf

ECLAC. (2017). Air Transport as a driver of sustainable development in Latin America and Caribbean: Challenges and policy proposals. *Bulletin FAL, 7*(259). https://repositorio.cepal.org/bitstream/handle/11362/43412/S1800005_en.pdf

Eklington, J. (1998). *Cannibals with forks: The Triple Bottom Line of the 21st century.* Stoney Creek, CT: New Society Publishers.

El-Takriti, S., Pavlenko, N., & Searle, S. (2017). *Mitigating international aviation emissions: Risks and opportunities for alternative jet fuels.* White paper, International Council on Clean Transportation.

Filippone, A. (2017). Options for aircraft noise reduction on arrival and landing. *Aerospace Science and Technology, 60*, 31–38. doi:10.1016/j.ast.2016.10.027

Flores, R., Asensio, C., Gagliardi, P., & Licitra, G. (2019). Study of the correction factors for aircraft noise façade measurements. *Applied Acoustics, 145*, 399–407. doi:10.1016/j.apacoust.2018.10.007

Fuglestvedt, J., Berntsen, T., Myhre, G., Rypdal, K., & Skeie, R. B. (2008). Climate forcing from the transport sectors. *Proceedings of the National Academy of Sciences of the United States of America, 105*(2), 454–458. doi:10.1073/pnas.0702958104 PMID:18180450

Gagliardi, P., Teti, L., & Licitra, G. (2018). A statistical evaluation on flight operational characteristics affecting aircraft noise during take-off. *Applied Acoustics, 134*, 8–15. doi:10.1016/j.apacoust.2017.12.024

Gittens, A., Juniac, A., Pie, J., Poole, J., & Liu, F. (2017). *Aviation Benefits 2017.* ICAO. https://www.icao.int/sustainability/Documents/AVIATION-BENEFITS-2017-web.pdf

Glasgow Airport. (2014). *Noise action plan 2013-2018.* https://www.glasgowairport.com/media/2067/current-noise-action-plan.pdf

Gupta, M., & Sharma, K. (1996). Environmental operations management: An opportunity for improvement. *Production and Inventory Management Journal, 37*, 40–46.

Herdman, D. (2018). *Asia Pacific Aviation: Competitive dynamics & growth.* Willis Towers Watson Aviation Conference. http://www.aapairlines.org/wp-content/uploads/2018/04/AAPA_SP_Herdman_WillisTowersWatsonAviationInsurance_SIN_Mar18.pdf

Hondett, S. (2010). *Airline cabin waste recycling guide*. https://www.sustainableaviation.co.uk/wp-content/uploads/2018/06/Aircraft-Cabin-Waste-Recycling-Guide1.pdf

IATA. (2018). *Best industry practices for aircraft decommissioning (BIPAD)*. https://www.iata.org/publications/Documents/BIPAD.pdf

IATA. (2019a). *Improving environmental performance*. https://www.iata.org/whatwedo/environment/Pages/index.aspx

IATA. (2019b). *E-Ticketing*. IATA. https://www.iata.org/whatwedo/stb/Pages/e-ticketing.aspx

ICAO. (2015). *United States aviation greenhouses gas emissions*.https://www.icao.int/environmentalprotection/Lists/ActionPlan/Attachments/30/UnitedStates_Action_Plan-2015.pdf

ICAO. (2019). *Environmental protection*. http://www.ICAO.int/environmental-protection/pages/default.aspx

Jun, T. L., & Jie, C. J. (2018). Research on China Airport Performance Evaluation of Energy Conservation and Emission Reduction Based on Data Envelopment Analysis Model. In *IOP Conference Series: Earth and Environmental Science* (Vol. 189, No. 5, p. 052011). IOP Publishing. 10.1088/1755-1315/189/5/052011

Lee, D. S., Fahey, D. W., Forster, P. M., Newton, P. J., Wit, R. C., Lim, L. L., ... Sausen, R. (2009). Aviation and global climate change in the 21st century. *Atmospheric Environment*, *43*(22-23), 3520–3537. doi:10.1016/j.atmosenv.2009.04.024

Lee, K. C., Tsai, W. H., Yang, C. H., & Lin, Y. Z. (2018). An MCDM approach for selecting green aviation fleet program management strategies under multi-resource limitations. *Journal of Air Transport Management*, *68*, 76–85. doi:10.1016/j.jairtraman.2017.06.011

Li, X. D., Poon, C. S., Lee, S. C., Chung, S. S., & Luk, F. (2003). Waste reduction and recycling strategies for the in-flight services in the airline industry. *Resources, Conservation and Recycling*, *37*(2), 87–99. doi:10.1016/S0921-3449(02)00074-5

Liu, X., Zhou, D., Zhou, P., & Wang, Q. (2017). Dynamic carbon emission performance of Chinese airlines: A global Malmquist index analysis. *Journal of Air Transport Management*, *65*, 99–109. doi:10.1016/j.jairtraman.2017.09.009

London Stansted Airport. (2010). *Making a material difference: London Stansted Airport waste management strategy 2010-2015.* https://live-webadmin-media.s3.amazonaws.com/media/3414/waste-strategy-stansted.pdf

McCollum, D. L., Gould, G., & Greene, D. L. (2010). *Greenhouse gas emissions from aviation and marine transportation: Mitigation potential and policies.* White paper, Pew Centre on Global Climate Change.

Mehta, P. (2015). Aviation waste management: An insight. *International Journal of Environmental Sciences*, *6*(1), 179.

Migdadi, Y. K. (2011a). *Identifying the best practice in traditional and electronic banking operations strategies in developing economies (The case of Jordan). Balancing resources with customer expectations in rapidly developing business environment* (Doctoral dissertation). University of Bradford.

Migdadi, Y. K. A. (2012). The developing economies' banks branches operational strategy in the era of e-banking: The case of Jordan. *Journal of Emerging Technologies in Web Intelligence*, *4*(2), 189–197. doi:10.4304/jetwi.4.2.189-197

Migdadi, Y. K. A. A. (2011b). The Impact of Adopting E-banking on Branches Operations Strategy in Developing Economies: The Case of Jordan. *Information & Communication Systems*, 208.

Migdadi, Y. K. A. A. (2013). Proposed typologies of banking operations strategy. *International Journal of Services and Operations Management*, *16*(1), 42–85. doi:10.1504/IJSOM.2013.055557

Migdadi, Y. K. A. A. (2016). Identifying the best practices in green operations strategy of leading mobile phone producers. *International Journal of Business Excellence*, *9*(1), 92–112. doi:10.1504/IJBEX.2016.073377

Migdadi, Y. K. A. A. (2018). Identifying the best practices of airlines' green operations strategy: A cross-regional worldwide survey. *Environmental Quality Management*, *28*(1), 21–32. doi:10.1002/tqem.21575

Migdadi, Y. K. A. A. (2019). Identifying the effective taxonomies of airline green operations strategy. *Management of Environmental Quality*.

Migdadi, Y. K. A. A. (2020). Green Operational Strategy for Airlines: Content and Regional Analysis. In Cases on Green Energy and Sustainable Development (pp. 193-229). IGI Global.

Migdadi, Y. K. A. A., & Elzzqaibeh, D. A. S. I. (2018). The evaluation of green manufacturing strategies adopted by ISO 14001 certificate holders in Jordan. *International Journal of Productivity and Quality Management*, *23*(1), 90–109. doi:10.1504/IJPQM.2018.088610

Migdadi, Y. K. A. A., & Omari, A. A. (2019). Identifying the best practices in green operations strategy of hospitals. *Benchmarking: An International Journal*, *26*(4), 1106–1131. doi:10.1108/BIJ-09-2017-0242

Olawuy, D. S. (2016). *Sustainable aviation and the transfer of environmentally sound technologies in Africa: Paradox, barriers and prospects. Occasional paper No. XIII. McGrill University*. Quebec University. https://www.mcgill.ca/iasl/files/iasl/xiii_sustainable_aviation_and_the_transfer_of_environmentally_sound_technologies_final.pdf

Ortega Alba, S., & Manana, M. (2016). Energy research in airports: A review. *Energies*, *9*(5), 1–19. doi:10.3390/en9050349

Pretto, M., Giannattasio, P., De Gennaro, M., Zanon, A., & Kühnelt, H. (2019). Web data for computing real-world noise from civil aviation. *Transportation Research Part D, Transport and Environment*, *69*, 224–249. doi:10.1016/j.trd.2019.01.022

Reiss, M., & Becker, A. (2018). *Managing cabin waste: market analysis and future concept. Airbus cabin waste management*. https://www.passengerexperienceconference.com/RXUK/RXUK_PexConference/Hamburg%202018/documents/Airbus_Cabin_Waste_SolutionsPresentation_PEC_final_Handout.pdf?v=636609410168031406

Sameh, M. M., & Scavuzzi, J. (2016). *Sustainable development in the aviation industry in Latin America: focus on Brazil. Occasional paper No.IX, July. McGrill University*. Quebec University. https://www.mcgill.ca/iasl/files/iasl/xi_sustainable_development_initiatives_in_the_aviation_industry_in_latin_america_final.pdf

Shamanskyi, S., & Boichenko, S. (2016). Development of environmentally safe technological water disposal scheme of aviation enterprise. *Восточно-европейский журнал передовых технологий*, *6*(10), 49-57.

Sotos, M. (2015). *GHG Protocol Scope 2 Guidance. An amendment to the GHG protocol corporate Standard. World Resources Institute (WRI), in association with the World Business Council for Sustainable Development*. WBCSD.

S&P Dow Jones Indices & RobecoSAM. (2019). *Dow Jones Sustainability Indices Methodology*. https://us.spindices.com/documents/methodologies/methodology-dj sustainability-indices.pdf

Sukumaran, S., & Sudhakar, K. (2017). Fully solar powered Raja Bhoj International Airport: A feasibility study. *Resource-Efficient Technologies*, *3*(3), 309–316. doi:10.1016/j.reffit.2017.02.001

Tofalli, N., Loizia, P., & Zorpas, A. A. (2018). Passengers waste production during flights. *Environmental Science and Pollution Research International*, *25*(36), 35764–35775. doi:10.100711356-017-0800-x PMID:29264848

Torija, A. J., Roberts, S., Woodward, R., Flindell, I. H., McKenzie, A. R., & Self, R. H. (2019). On the assessment of subjective response to tonal content of contemporary aircraft noise. *Applied Acoustics*, *146*, 190–203. doi:10.1016/j.apacoust.2018.11.015

UN-Secretary-General-High-Level Advisory Group on Sustainable Transportation. (2015). *Mobilizing sustainable transport for development*. https://sustainabledevelopment.un.org/content/documents/2375Mobilizing%20Sustainable%20Transport.pdf

Wieher, C. R. (2014). *Hazardous Waste Curriculum for Aviation Maintenance*. Brevard Teaching Research and Laboratories: Florida. https://www.broward.org/EnvironmentAndGrowth/EnvironmentalProgramsResources/Publications/Documents/CurriculumAviationMaint.pdf

Wyman, O. (2012). *Guide to airport performance measures. Airports Council international*. ACI.

Xu, J. K., & Deng, Y. W. (2018). Research on Application of Renewable and Clean Energy in Beijing New Airport. *DEStech Transactions on Environment, Energy and Earth Sciences*.

Yan, W., Cui, Z., & Gil, M. J. Á. (2016). Assessing the impact of environmental innovation in the airline industry: An empirical study of emerging market economies. *Environmental Innovation and Societal Transitions*, *21*, 80–94. doi:10.1016/j.eist.2016.04.001

Chapter 2
Airline Effective Direct Greenhouse Gases Emissions Strategies

ABSTRACT

This chapter reports the effective direct GHG emissions strategies. To achieve this aim, the flight and the ground support practices of 22 airlines from five regions were investigated. The data used in this chapter was secondary published data in the annual sustainability reports. The data was retrieved from GRI (Global Reporting Initiatives) website. The special tailored data analysis techniques were used for this purpose. The chapter reported eight effective flight operations strategic patterns and two effective ground operations strategic patterns. The strategic patterns of flight operations were two design+ patterns, three rout+ patterns, two operating+ patterns, and one fuel pattern. However, the effective ground operations strategies were two design+ patterns. This chapter helps decision makers and academics alike, since the impact of each action on reducing the direct GHG emissions was reported.

DOI: 10.4018/978-1-7998-4255-2.ch002

INTRODUCTION

The aviation is responsible of 12% of CO2 emissions from all transportation sectors and it produces about 2% of all human induced carbon dioxide (CO2) emissions (Air Transport Action Group, 2018). The environmental impact of airlines has increased over the last 25 years because of increase in demand (Lee *et al.,* 2009). CO2 emissions of airlines are expected to increase by further 45% between 2014 and 2035. The future green targets of aviation as adopted by IATA are improving the average efficiency in fuel by 1.5% per year from 2009 to 2020, and reducing the net aviation CO2 emissions of 50% by 2050 relative to 2005 level. (IATA, 2019a). Accordingly, the aviation industry is responsible about making more actions to reduce the negative environmental impact. All stakeholders must make a genuine commitment to transform the transport system toward ecological friendly practices (UN-Secretary-General-High-Level Advisory Group on Sustainable Transportation, 2015).

In this regard, the airlines should take more effective actions to reduce GHG emissions. The actions taken by airlines are diverse and numerous, but the focus should be on the most effective actions to be adopted. The direct GHG emissions that generated from flight and ground support operations represent the significant portion of total GHG emissions. Accordingly, the airlines should have a better insight about the most effective actions to reduce the direct GHG emissions. The previous studies have investigated the green actions that will reduce GHG emissions. The most of previous studies have focused on reporting the flight operations actions (e.g. Chapman, 2016; El-Takriti, Pavlenko & Searle, 2017; Lin, Zahou, Zahou & Wang, 2017; Lee, Tsai, Yang & Lin, 2018). The actions have been traced by most studies were limited in their scope. Most of previous studies have reported the actions with limited concern about the different strategic patterns were adopted.

This chapter bridged the gap of previous studies by reporting the effective green operations strategies in reducing direct GHG emissions. This chapter reported the practices and effective strategic patterns of many airlines in different regions. This chapter adopted the benchmarking methodology by comparing the performance of airlines with the more effective airline, and reporting the effective strategy patterns. Accordingly, the main question of this chapter was: what were the most effective green strategic patterns adopted by airlines to reduce the direct GHG emissions? To answer the question the following objectives were realized:

1. Ranking airlines according to the achieved direct GHG emissions performance.
2. Identifying the effective direct GHG emissions strategic patterns that adopted by airlines.
3. Identifying the significant green actions taken by the effective strategic patterns.
4. Identifying the impact of significant green actions on reducing the direct GHG emissions.

This chapter is arranged in five sections: the first one is the literature review, followed by the methodology and methods, then, the results of the empirical study, next, the discussion, the final section is the conclusions.

LITERATURE REVIEW

This part of the chapter discussed the key literature of airlines direct GHG emissions. The first part of the literature aimed to discuss how to define and measure the direct GHG emissions, so different methodologies and models of measuring the direct GHG emissions were presented, also the factors and assumptions that should be considered in measuring the direct GHG emissions were discussed. The second part of the literature review aimed to discuss the green operations actions, and the different strategic patterns to reduce direct GHG emissions.

AIRLINE DIRECT GHG EMISSIONS MEASUREMENT

The airlines greenhouse gas emissions include the emissions of carbon dioxide CO_2, the emissions of NOx, emissions of sulphate arising from sulphur in fuel, water vapor and formation of cirrus clouds (Bofinger & Strand, 2013). These emissions could be classified according to its effect on the climate into direct emissions, indirect emissions, emissions of aerosols, and formation of contrails and cirrus clouds. The direct GHG emissions according to the previous classification includes such emissions as CO_2 and water vapor. However, the indirect GHG emissions includes NOx emissions. The emissions of aerosols or the emissions have cooling or warming effect as sulfates and soot. The formation of contrails and cirrus clouds, which are emissions occur in the troposphere and lower stratosphere (Fuglestvedt et al., 2008).

Another direct and indirect emissions classification scheme of airlines has adopted different definitions. For example, the direct emissions or what is called GHG1 emissions are defined as the metric tons of CO_2 or equivalent CO_2 emissions of aircrafts and ground support operations. However, the indirect emissions could be classified into GHG2 and GHG3. GHG 2 emissions are defined as metric tons of indirect CO_2 emissions of electricity, power and heat from the direct billing of owned or leased facilities. However, GHG3 emissions are defined as metric tons of indirect CO_2 emissions, that are emitted from any parts of the value chain out of GHG 2 emissions (Migdadi, 2018; Migdadi, 2020).

Different protocols have been developed to calculate the GHG emissions of aviation sector. For example, DEFRA (Department of Environmental Food and Rural Affairs) protocol in the UK. ICAO (International Civil Air Transport Organization) protocol. Each protocol has its own assumption and methodology. DEFRA calculates the emissions per passenger based on load factors of 66.3%, 81.2% and 78.1% for domestic, short-haul and long-haul respectively. The emissions are allocated between economy and premium classes. However, ICAO calculated per passenger emissions based on load factor. The calculated emissions vary based on route group and whether plan is wide or narrow body (Jardine, 2009).

Different models have been used to estimate the impact of aviation. Some of these models calculate how the greenhouse gases emissions change the energy balance in the atmosphere. Other types of models calculate the temperature change that will be caused by the change in energy balance. By combining two types of models, a more comprehensive measurement methodology could be adopted. Accordingly, it is required to calculate the greenhouse gases emissions from human activities. Then, calculating the atmospheric greenhouses gases concentrations. Next, determine radiative forcing, modeling climate responses as climate feedback, climate efficacy and climate sensitivity. Followed by modeling the climate impacts and modeling the climate damages or costs (Kollmuss & Crimmins, 2009).

The methodology adopted in calculating the aviation emissions will affect the uncertainty of these calculations. Also, much uncertainty comes from the fact that emissions at altitude depends on chemical reactions on the atmosphere, which each have global warming and cooling effect over a variety of timescales (Jardin, 2009). Different factors affect per passenger emissions. These factors includes the plan type, the engine type of the plan, the seating configuration, and the flight load (Jardine, 2009). The calculating of the average GHG emissions per traveled distance is affected by many factors as the class of travel, the load factor, the weight factor, and other factors as the flight length and the average altitude (Bofinger & Strand, 2013). Accordingly, aviation emissions calculators should have clear assumptions about the previous factors (Jardine, 2009).

Airlines Direct GHG Emissions Operational Actions and Strategy Patterns

The airline direct GHG emissions operational actions are those actions related to flight and ground supporting actions (Migdadi, 2018). These actions are classified into flight operations and ground operations. The flight operations actions are classified into different categories as the technological advancements, the operational improvements, and the alternative fuels (El-Takriti, Pavlenko & Searle, 2017). Another classification scheme classified actions into operations actions, the aircraft design and the alternative fuel (McCollum, Gloud & Greene, 2009). Another classification scheme combined them under two main categories as technology and process-based (Yan, Cui & Gil, 2016).

The previous classifications have focused on flight operations. However, the most comprehensive and extended classification has been developed by Migdadi (2018; 2020). The flight operations have been classified into the aircraft design, the route management, the flight operating, the flight weight management, the flight fuel management and the flight maintenance management. However, the ground operating actions categories have been classified into the vehicles and engines design, the operating, the maintenance and the transportation management (Migdadi, 2018; 2020). Under each category, there is many listed actions.

The impact of actions categories in reducing GHG emissions differs. The fuel saving actions were the most significant in reducing direct GHG emissions (Migdadi, 2018). Another study found that CO2 emissions performance of aviation is influenced by route distribution, followed by fuel consumption rate, aircraft utilization rate, and movement (Lin, Zahou, Zahou & Wang, 2017). The study of Abdullah, Chew & Hamid (2016) confirms the previous results. The most effective actions category was the daily operational actions as fuel management, aircraft weight reduction, flight planning and ground operations. However, the least effective actions category was corporate policy and strategies as fleet renewal. In addition, this result was confirmed by El-Takriti, Pavlenko & Searle (2017). The first ranked actions category in reducing GHG emissions was fuel and market based.

The impact of actions and sub-actions categories differs. For example, the rank of operating actions differs. The reducing of APU usage is the first ranked followed by the weight reduction, then, the air-traffic management, next, the flight procedures, the route optimization and the alternative bio-fuel (Lee, Tsai, Yang & Lin, 2018). However, the operating of larger aircraft as B737-700 with blended winglets reduces the fuel consumption and CO2 emissions by 2.7%. Nevertheless, engine-washing programs reduces CO2 emissions by 1.2% (Chapman, 2016). The operations of larger aircraft is beneficial in reducing the fuel usage and the carbon emissions (Brueckuer & Abreu, 2017; Ma, Song & Zhu, 2018). Accordingly, the charter type carriers with a high-density interior configuration, medium and long-range routes, high load factor and larger aircrafts are the most efficient CO2 emissions (Amizadeh, Alonso, Benito & Morales-Alonso, 2016). On the other hand, the fuel consumption per seat during LTO (Landing, Take-off cycle) for smaller with single-aisle (narrow body) jets is lower than larger with double-aisle (wide body). Also, the cost the LAP (Local Air Pollution) (Euro/seat) of emissions is lower for smaller than larger. Moreover, the engine thrust during take-off and climb is lower for smaller aircrafts. In general based on fuel consumption, larger aircrafts appear to have lower environmental impact on climate changes per seat than smaller (Givoni & Rietveld, 2010).

The most of previous studies have investigated the impact of flight operations in reducing GHG emissions. However, the impact of ground operations is rarely investigated. According to Migdadi (2018), the effective action category in reducing GHG emissions was the vehicles and the engine operating action. However, this study has not investigated the impact of actions under the action categories. On the other hand, the GHG strategic patterns adopted by airlines to reduce GHG emissions are rarely investigated. The study of Migdadi (2018) reported the general green strategic patterns but not investigated in details the effective strategic patterns in reducing direct GHG emissions.

METHODOLOGY AND METHODS

This part of the chapter discussed how the strategic patterns were identified, and how the effectiveness of these patterns was assessed. In this regards, the design of sample was discussed. In addition, how the data was collected and from which sources was discussed. Then, how the data was analyzed to formulate strategy patterns, and track the effectiveness of these patterns was discussed in details. Finally, supportive examples on data analysis approach were presented in the appendices at the end of the chapter.

Sample Design

The sample of study was a convenient sample according to availability of secondary published data. Table 1 shows the distribution of the sample between airline operations and regions. It can be seen that the flight operations were surveyed for 22 airlines. The ground operations were surveyed for 8 airlines. Table1 shows the distribution of sample between regions. It can be seen that the flight operations of six airlines from each region of North America, Asia and Europe were analyzed. Three airlines from the Middle East and one airline from the South America were surveyed. The ground operations of three airlines from each region of North America and Europe were analyzed. One airline from each region of Asia and the Middle East was analyzed.

Table 1. Sample of study

Regions	Flight Operations	Ground Operations
NA: North America	6	3
A: Asia	6	1
E: Europe	6	3
ME: Middle East	3	1
SA: South America	1	0
Total	22	8

Sources of Data

The data was collected from annual sustainability report of airlines. The reports were retrieved from Global Reporting Initiatives (GRI) database. The GRI is an independent international organization established in 1997 in the US. It seeks to help organizations to report their impact on sustainability. It develops standards of reporting sustainability and publishes reports via its website (GRI, 2017).

Definition and Measures of Variables

The dependent variable was the direct GHG emissions (GHG1). However, the independent variables were the flight and ground operations actions. Direct GHG emissions (GHG1) were defined as metric tons of direct CO_2 emissions from the jet fuel and the ground support operations. However, the flight operations actions were classified into different dimensions as the aircraft design, the rout management, the flight operating, the flight weight management, the flight fuel management and the maintenance management. Each dimension of flight operations actions was measured by set of actions. The list of actions is presented in Table 2. In addition, the ground operations actions were classified into different dimensions as the vehicles and engines design, the vehicles and the engines operating, the vehicles and the engines maintenance management and the transportation management. Each dimension of ground operations actions was investigated by set of actions. The list of actions is presented in Table 3.

Data Analysis

How to assess the direct GHG emissions performance

Phase1: Computing the percentage of change in direct GHG emissions. The following steps were taken to achieve this phase:

1. Collecting the actual achieved performance in 2013 and 2016.
2. Computing the percentage of change by using the following formula

[(GHG emissions 2016- GHG emissions 2013)/ GHG emissions 2013] X100%

Phase 2: Transforming the percentage of change in direct GHG emissions into rating value as presenting in Table 4. The following steps were taken to achieve this phase:

1. Identifying the case achieved the best performance
2. Dividing the performance achieved by each case by the best, the result is the relative performance
3. Multiplying the relative performance by 5 (the maximum rating value of the scale). The result is the rating value of performance.

Table 2. Direct GHG Emissions Flight Operations Actions

Actions Dimensions	Actions
Aircraft Design	Fleet modernization
	Split scimitar winglets
	Shark lets wings
	Engine modification
Route Management	Routes optimization
	PC calculate flight path
	Broadcasting technology for communicate everywhere
	Monitoring actual flight path and relevant data
	Real time data text communication
	Using GPS
	Center of gravity
Flight Operating	Reduce use of APU
	Single taxiing engine
	Acceleration altitude
	Accurate planning of flight
	Altitude optimization
	Reducing aircrafts take-off time
	Reducing flight launching time
	Reducing use of extended landing lights
	Reducing landing time
	Optimal using of flap
	Reduce using of idle reverse thrust when landing
	Reducing aircraft waiting time
	Landing continuous descent operations
	Reducing taxing time
	Revise and improve fuel loading regulations
Flight Weight management	Optimal water potable carried
	Using EFB
	Using on board tablets
	Light meal trolley
	Eliminate magazines
	Digital route charts
	Lighter seats
	Light weight cargo containers
	Installing zonal driers in aircrafts as dehumidifiers which reduce moister weight
	Restrict fuel overload
	Reducing number of flight operations manual onboard
	Using lighter galley
	Reduce pay load variance
Flight fuel management	Using bio-fuels
	Training on fuel saving
	Installing fuel management and reporting system
	Reducing contingency fuel planning for long haul routes
Maintenance management	Engine washing
	Maintain interior aircraft plumbing
	Aircraft body washing
	Resort the exterior of fan blades in the engines

(Source: Migdadi, 2018, 2020)

Table 3. Direct GHG Emissions Ground Operations Actions

Actions Dimensions	Actions
Vehicle and engine design	Electrical carts or tractor (trucks) equipment
	Using battery vehicles for ground operations
	Using renewable energy-fueled equipment
Vehicle and engine operating	Reducing vehicles standby time
	Management of office vehicle trips
	New forklifts
	Full more cargo at once
Vehicle and engine maintenance management	Engines water wash
	Cleaning vehicles and equipment
Transportation management	Cars sharing
	Low CO2 emissions cars
	Subsided shuttle bus
	Pool bicycles
	Electrical vehicles charging points

(Source: Migdadi, 2018, 2020)

Please see the example in the Appendix, Table 17.

Phase 3: Ranking the cases according to achieved rating value. The following steps were taken to achieve this phase:

1. Classifying the scale into three rating levels as following:

1. Step 1: identifying the turning value as following

[(5-0)]/3= 1.67

Step 2: Identifying the levels by adding the turning value as following: From more than 0 to 1.67 is low level. From more than1.67 to 3.33 is moderate level. From more than 3.33to 5.00 is high level.

Table 4. Rating Scale of the Percentage of Change in Direct GHG Emissions

0	1	2	3	4	5
	Extremely low	Low	Moderate	High	Extremely high

2. Assign the level for each case according to achieved rating value. The Assigned level represents the rank of the case performance.

Please see the example in the Appendix, Table 17.
How to identify the effective direct GHG emissions strategic patterns

Phase 1: Computing the actions indices of direct GHG emissions. The following steps were taken to achieve this phase:

1. Identifying the actions categories and actions lists. This study adopted the same classification of Migdadi (2018, 2020).
2. Computing the number of taken actions by each case under each category.
3. Dividing the number of taken actions by each case by the total number of actions could be taken. The result is a relative number, which is called the action adoption index.

Please see the example in the Appendix, Table 18.

Table 5. Rating Scale of Reducing Direct GHG Emissions Actions Index

0	1	2	3	4	5
	Extremely low	Low	Moderate	High	Extremely high

Phase 2: Transforming the action taken index into rating value as presented in Table 5. The following steps were taken to achieve this phase:

1. Identifying the case achieved the best index value.
2. Dividing the action index achieved by each case by the best. The result is the relative action taken index.
3. Multiplying the relative index value by 5 (the maximum rating value of the scale). The result is the rating value of actions taken.

Please see the example in the Appendix, Table 19.

Phase 3: Identifying the relative level of action adoption according to achieved rating value. The following steps were taken to achieve this phase:

1. Classifying the scale into three rating levels as following:

1. Step 1: identifying the turning value as following

[(5-0)]/3= 1.67

Step 2: Identifying the levels by adding the turning value as following: From more than 0 to 1.67 is low level. From more than1.67 to 3.33 is moderate level. From more than 3.33to 5.00 is high level.

2. Assign the level for each case according to achieved rating value.

Phase 4: Summarizing the actions categories indices for all cases. The cases were ordered according to ranking value of achieved direct GHG emissions.

Phase 5: Identifying the actions categories with moderate and high rating value.

Phase 6: Identifying the cases that took shared actions categories with moderate or high level rating value.

Phase 7: Assigning the titles for the cases according to most shared actions between cases. These titles represent different adopted patterns. If the patterns were adopted by the cases that achieved an improvement in performance, then, the patterns would be recognized as the effective patterns.

Please see the example in the Appendix, Tables 20 through 24.

How to compute the impact of actions taken by the effective patterns on the achieved direct GHG emissions

Phase 1: Identifying the actions taken by the higher effective cases in comparison with the lower effective and non-effective cases (which adopted the same pattern). These actions were called the significant actions. Please see the example in the Appendix, Table 25.

Phase 2: Counting the number of significant actions taken under each action category by the higher effective cases. Please see the example in the Appendix, Table 26.

Phase 3: Computing the relative impact of each action category by dividing the number of significant actions under each category by the total number of significant actions. Please see the example in the Appendix, Table 27.

Phase 4: Computing the differences in GHG emissions between the higher effective and the lower effective or non-effective case as following:

Direct GHG emission of higher effective case - direct GHG emission of lower (or non-effective) case (Please see the example in the Appendix, Table 28).

Phase 5: Multiplying the relative impact of each action category by the differences in GHG emissions between the higher and the lower effective or non-effective case. The result is the reduction in GHG emissions achieved by the action category. Please see the example in the Appendix, Table 29.

Phase 6: Dividing the reduction in GHG emissions achieved under the action category by the number of actions of the category. The result is the amount of reduction in GHG emissions achieved by the action.

Phase 7: Computing the average reduction of GHG emissions achieved by each action. The result is the average impact of the action under each category.

Please see the Example in the Appendix, Table 30.

RESEARCH RESULTS

This part of this chapter presented the main results of the empirical study. The aim of this part was to identify the effective green operations strategies for the flight and the ground operations. This part started by presenting the results related to the flight operations followed by the ground operations. The ranking of airlines according to the achieved direct GHG emissions of the flight and the ground operations were presented first. Then, the strategic patterns adopted according to realized actions indices were presented. Finally, the strategic patterns of the flight and the ground operations were presented as indicative models. The indicative models include the impact of actions adopted by each strategic pattern on the reduced direct GHG emissions. In order to facilitate the reading of this chapter results Table 6 shows the acronyms of the strategy patterns.

Table 6. List of Acronyms Used in Presenting the Chapter Results

Acronym	Interpretation
D	Aircraft Design
F	Flights fuel management
GHG1	Direct greenhouses gas emission
GHG2	Indirect greenhouse gas emission
M	Flights maintenance management
O	Flight Operating
R	Flights routes management
VD	Engine and vehicle design
VO	Engine and vehicle Operating
VM	Engine and vehicle maintenance
W	Flights weight management

Direct GHG Emissions of Airlines Flight Operations

Table 7 shows the direct GHG emissions of airlines flight operations. It can be seen that 50% (11 cases) of the cases reduced the direct GHG emissions. The percentages of reduction in the direct GHG emissions was between -3% to -21%. The level of reduction according to the rating scale levels was low for nine cases, which was about 82% of the reduced emissions cases. However, two cases achieved a high reduction in the direct GHG emissions. The percentage of increase in the direct GHG emissions was between 4% to 34%. The level of increase according to the rating scale level was low for seven cases, which was about 64% of the increased emissions cases. Three cases achieved a moderate increase in the direct GHG emissions, which was about 18% of the increased emissions cases and one case achieved a high increase in the direct GHG emissions.

Direct GHG Emissions of Airlines Ground Operations

Table 8 shows the direct GHG emissions of airlines of ground operations. It can be seen that 50% (four cases) of the cases reduced the direct GHG emissions, and the remaining 50% of the cases increased the direct GHG emissions. The percentages of reduction in the direct GHG emissions was between -2.12% to -34.70%. The level of reduction according to the rating scale levels was low by two reduced emissions cases. However, one case achieved moderate

Table 7. Direct GHG Emissions of Airlines Flight Operations

Case ID	% of change in emissions	5-points rating of change in emissions	Rating level
Reduced Emissions Cases			
E3	-21%	+5.0	+H
ME3	-20%	+4.7	+H
E6	-7%	+1.6	+L
A5	-5%	+1.2	+L
E5	-4%	+0.9	+L
NA5	-2%	+0.6	+L
E1	-2%	+0.5	+L
SA1	-2%	+0.5	+L
A7	-2%	+0.4	+L
NA2	-2%	+0.4	+L
Increased Emissions Cases			
A6	+0.3%	-0.1	-L
NA1	+4%	-0.6	-L
A1	+4%	-0.6	-L
E2	+5%	-0.7	-L
E4	+5%	-0.8	-L
NA3	+9%	-1.4	-L
A3	+9%	-1.4	-L
A4	+11%	-1.6	-L
NA6	+13%	-1.9	-M
NA4	+16%	-2.4	-M
ME2	+20%	-3.0	-M
ME1	+34%	-5.0	-H

L: Low **M:** Moderate **H:** High
+: Improvement in the performance (Reduction in the direct GHG emissions)
-: Deterioration in the performance (Increase in the direct GHG emissions)

reduction in the direct GHG emissions and one case achieved a high reduction in the direct GHG emissions. The percentage of increase in the direct GHG emissions was between 1.2% to -3.0%. The level of increase according to the rating scale level was low by four cases and moderate by one case.

Table 8. Direct GHG Emissions of Airlines Ground Operations

Case ID	% of change in emissions	5-points rating of change in emissions	Rating level
Reduced Emissions Cases			
E2	-34.70%	+4.7	+H
E3	-21.14%	+2.9	+M
NA3	-8.62%	+1.2	+L
E1	-2.12%	+0.3	+L
Increased Emissions Cases			
ME1	8.68%	-1.2	-L
A3	11.61%	-1.6	-L
NA6	12.91%	-1.7	-L
NA5	22.35%	-3.0	-M

L: Low **M:** Moderate **H:** High
+: Improvement in the performance (Reduction in the direct GHG emissions)
-: Deterioration in the performance (Increase in the direct GHG emissions)

Effective Direct GHG Emissions Strategic Patterns of Flight Operations

Table 9 Shows the actions indices and the strategy patterns of flight operations. It can be seen that the levels of realized actions indices were low, moderate and high. For the purpose of this study, only the moderate and high-level actions indices were recognized as the significant actions taken by airlines. It can be seen that some cases took a combination of different significant actions as ME3, NA5 and A7. Other cases took only one action as E3, E6 and ME1. On the other hand, the significant actions were taken by the reduced emissions and the increased emissions cases. The significant actions taken by the surveyed cases were used as a source to identify the strategic patterns of the cases. Accordingly, the last column of Table 9 shows these patterns. Some patterns were more closer than others in term of the shared significant actions. Each pattern has a title according to the most shared actions between the close patterns cases. The patterns adopted by airlines were fuel pattern, which was adopted by two cases, one of them was of the increased emissions cases. Design + pattern, which was adopted by six cases, four of them were of the increased emissions cases. Maintenance pattern, which was adopted by one case of the reduced emissions cases. Rout + pattern, which was adopted by seven cases, four of them were of the increased emissions cases.

Table 9. Actions Indices and Strategy Patterns of Flight Operations

level of % of change in Direct GHG emissions	Case ID	Aircraft design (D)	Flights routes management (R)	Flights Operating (O)	Flights weight management (W)	Flights fuel management (F)	Flights maintenance management (M)	Strategy Pattern	Strategy pattern title
Reduced Emissions Cases									
+H	E3	0.25 (L)	0.14 (L)	0.2 (L)	0.15 (L)	0.5 (H)	0	F	Fuel
+H	ME3	0.75 (H)	0.57 (H)	0.33 (H)	0.54 (H)	0.5 (H)	0.25 (M)	D-R-O-W-F-M	Design +
+L	E6	0	0	0.13 (L)	0.08 (L)	0	0.25 (M)	M	Maintenance
+L	A5	0	0	0.27 (M)	0	0	0	O	Operating +
+L	E5	0.25 (L)	0.29 (M)	0.27 (M)	0.23 (M)	0	0	R-O-W	Rout +
+L	NA5	0.25 (L)	0.29 (M)	0.2 (L)	0.08 (L)	0.25 (M)	0	R-F	Rout +
+L	E1	0.25 (L)	0	0.4 (H)	0.15 (L)	0.25 (M)	0	O-F	Operating +
+L	SA1	0.25 (L)	0.57 (H)	0.27 (M)	0.15 (L)	0	0.25 (M)	R-O-M	Rout +
+L	A7	0.5 (M)	0.29 (M)	0.07 (L)	0	0	0.25 (M)	D-R-M	Design +
+L	NA2	0.25 (L)	0	0	0	0	0	-------	--------
Increased Emissions Cases									
-L	A6	0.25 (L)	0.14 (L)	0.27 (M)	0.3 (M)	0.5 (H)	0.5 (H)	O-W-F-M	Operating +
-L	NA3	0	0.25 (M)	0.07 (L)	0.23 (M)	0	0	R-W	Rout +
-L	A3	0	0.29 (M)	0.13 (L)	0.15 (L)	0	0.25 (M)	R-M	Rout +
-L	A4	0.25 (L)	0.43 (M)	0.47 (H)	0.23 (M)	0	0.5 (H)	R-O-W-M	Rout +
-L	NA6	0.5 (M)	0.14 (L)	0	0.31 (M)	0.25 (M)	0.25 (M)	D-W-F-M	Design +
-L	NA4	0.5 (M)	0.43 (M)	0.07 (L)	0	0	0	D-R	Design +
-L	ME2	0.5 (M)	0	0.13 (L)	0.15 (L)	0.25 (M)	0	D-F	Design +
-L	ME1	0	0.29 (M)	0.13 (L)	0.15 (L)	0.5 (H)	0	R-F	Rout +
-M	A1	0.5 (M)	0.14 (L)	0.07 (L)	0.15 (L)	0	0	D	Design+
-M	E2	0.25 (L)	0	0.13 (L)	0.08 (L)	0	0	-------	-------
-M	E4	0	0	0.13 (L)	0	0	0	------	-------
-H	NA1	0.25 (L)	0.14 (L)	0.07 (L)	0.15 (L)	0.25 (M)	0	F	Fuel

(L): Low (M): Moderate (H): High

Operating + pattern, which was adopted by three cases, one of them was of the increased emissions cases.

Effective Design$^+$ Strategic Patterns of Flight Operations

Table 10 shows the effective design+ strategic patterns. Two patterns were adopted. The first pattern was adopted by the case ME3. This pattern includes the actions categories of the design, the route, the operating and the weight. The second pattern was adopted by the case A7. The second pattern includes the actions categories of design, route and maintenance. Two shared actions categories between the two strategic patterns. These categories were design and route. The first pattern took more actions categories than the second pattern. The impact of each action category in reducing the direct GHG emissions of the flight operations varies for the first pattern (which was adopted by ME3). The most effective action category was weight by -17.2%. The second effective action category was route by -10.5%. The third effective action category was design by -5.2%. The fourth effective action category was operating by -5%. The lowest effective action category was maintenance by -2.3%. The impact of actions category in reducing the direct GHG emission for the second strategic pattern (which was adopted by A7) also varies. The most effective action category was the route management by -10.2%, followed by the design category with an impact by -5.5%, and the lowest effective action category was the maintenance management by an impact of -4.8%. The two patterns sharing the same ranking of the shared action categories in term of reducing the direct GHG emissions. However, the impact of the second strategic patterns actions categories was more than the first pattern.

The number of actions taken by the first strategy pattern varies. The largest number was related to the aircraft weight action categories by seven actions. Four actions related to the route management category. The lowest number of actions was related to the design and the operating by two actions for each category. The design actions taken by ME3 were shark let wings tips and split scimitar winglets. The impact of these two actions was -2.4% and -2.8% respectively. The actions of route category could be classified into technology utilization and process related. The technology utilization actions were PC-calculated flight path and broadcasting technology for communication everywhere. However, the process actions were the route optimization and monitoring the actual flight path and relevant data. The impact of process related action was more than the technology oriented. The impact was -2.4% for each technology oriented action. The impact of process-oriented actions was -2.7% for monitoring the actual flight path and relevant data, and -3% for the route optimization. Therefore, the route optimization is the most effective route action in reducing the direct GHG emissions.

The impact of operating related actions of the first strategy pattern (which was adopted by ME3) was close. The impact of acceleration altitude and the accurate planning of flight was -2.5% for each one. The actions of weight category could be classified into sub-categories as the replacing facilities with technology, (as using EFB and using on-board tablets), and using light weight facilities (as optimal water potable carried, light meal trolleys, light weight cargo containers and restricting fuel overload). The impact of lightweight facilities actions together was more than the replacing of facilities with technologies. The impact of each action of lightweight category was somehow close, but the most effective action was restricting fuel overload by an impact of -3%. The impact of remaining actions of lightweight category was between -2% to -2.5%. The lowest effective action was carrying the optimal potable water by an impact of -2%. The impact of each actions of replacing technology was the same for both actions (as using EFB (Electronic Flight

Bag) and using onboard tablets) by an impact of -2.5%. One action related to the maintenance category was taken by the case ME3. This action was engine washing and its impact was -2.3%. The total impact of the first strategy pattern (which was adopted by ME3) was -37.3%, which was more than the total impact of the second strategic pattern (which was adopted by A7).

The number of actions related to the route category (which was taken by A7) was two. The first action was the route optimization, but the second action was the monitoring of actual flight path and relevant data. The impact of these actions was -5.3% and -4.9% respectively. These two actions were taken by ME3 and the ranking of impact was the same. The route optimization impact was more than monitoring the actual flight path and relevant data. One deign action was taken by A7. This action was engine modification, its impact was -5.3%. Only one maintenance action was taken by A7. This action was the engine washing, and its impact in reducing the direct GHG emissions was -4.8%. The maintenance action taken by A7 was the same the action taken by ME3.

Effective Fuel Management Strategic Patterns of Flight Operations

Table 11 shows the effective fuel management strategic pattern. Only one pattern was adopted by the case E3. In addition, one action was taken by E3, this action was training in the fuel saving. The impact of this action in reducing the direct GHG emissions was -25%.

Effective Operating⁺ Strategic Patterns of Flight Operations

Table 12 shows the effective Operating+ strategic patterns. Two patterns were adopted. The first pattern was adopted by the case A5. This pattern includes the actions categories of operating. The second pattern which was adopted by the case E1. The second pattern includes the actions categories of the operating and the fuel. One shared actions category between the two strategic patterns, which was the operating category. The case of second pattern took more actions categories than the first pattern. The impact of each action category in reducing the direct GHG emissions of flight operations was equal for the first pattern (which was adopted by A5). The impact of each action category was -2.7%. The impact of the actions category in reducing the

Table 10. Effective Design+ Strategic Patterns of Flight Operations

Case ID	Strategy Pattern	Actions category Total Impact	Significant actions taken	Average Relative % of change in Direct GHG emissions
ME 3	D-R-O-W-F-M	**Design (D)** Total Impact= -5.2%	Shark let wing tips	-2.4%
			Split scimitar winglets	-2.8%
		Route (R) Total Impact= -10.5%	PC-calculated flight path	-2.4%
			Broadcasting technology for communication everywhere	-2.4%
			Route optimization	-3%
			Monitoring actual flight path and relevant data	-2.7%
		Operating(O) Total Impact= -5%	Acceleration altitude	-2.5%
			Accurate planning of flight	-2.5%
		Weight (W) Total Impact= -17.2%	Optimal water potable carried	-2%
			Using EFB (Electronic Flight Bag)	-2.5%
			Using onboard tablets	-2.5%
			Light meal trolleys	-2.3%
			Eliminating magazines	-2.4%
			Light weight cargo containers	-2.5%
			Restricting fuel overload	-3%
		Maintenance (M) Total Impact= 2.3%	Engine washing	-2.3%
Total Impact of the Pattern				-37.3%
A7	D-R-M	**Design (D)** Total Impact= -5.5%	Engine modification	-5.5%
		Route (R) Total Impact= -10.2%	Route optimization	-5.3%
			Monitoring actual flight path and relevant data	-4.9%
		Maintenance (M) Total Impact= -4.8%	Engine washing	-4.8%
Total Impact of the Pattern				-15.3%

Table 11. Effective Fuel Management Strategic Pattern of Flight Operations

Case ID	Strategy pattern	Action category Total Impact	Significant actions taken	Average Relative % of change in Direct GHG emissions
E3	F	**Fuel (F)** Total Impact= -25%	Training in fuel saving	-25%

Table 12. Effective Operating+ Strategic Patterns of Flight Operations

Case ID	Strategy pattern	Action category Total Impact	Significant actions taken	Average Relative % of change in Direct GHG emissions
A5	O	**Operating (O)** Total Impact= -5.4%	Reduce using of auxiliary power unit	-2.7%
			Landing continuous descent operations	-2.7%
Total impact of the pattern				-5.4%
E1	O-F	**Operating (O)** Total Impact= -2%	Reduce using of auxiliary power unit	-0.5%
			Reducing aircraft take off time	-0.5%
			Reducing flight launching time	-0.5%
			Reduced taxiing time	-0.5%
		Fuel (F) Total Impact= -0.4%	Using biofuels	-0.4%
Total impact of the pattern				-2.4%

Note: No significant actions related to maintenance (M) were taken by A5 and E1 in comparison with non-effective case A6

direct GHG emissions for the second strategic pattern (which was adopted by E1) varies. The most effective action category was operating by an impact of -2%, followed by the fuel category with an impact of -0.4%.

Two operating actions were taken by the case A5. The first action was reduce using of the auxiliary power unit. However, the second action was landing continuous descent operations. The impact of each action on reducing the direct GHG emission was -2.7%. The total impact of the first strategy pattern (which was adopted by A5) was -5.4%, which was more than the total impact of the second strategic pattern (which was adopted by E1). Four actions related to operating category were taken by the second case E1. These actions were reduce using of the auxiliary power unit, reducing the aircraft take off time, reducing the flight launching time and reducing the taxiing time. The impact of the previous four actions was equal by an impact of -0.5%. One action related to the fuel management was taken by E1. This action was using biofuels. The impact of this action was -0.4%.

Effective Route+ Strategic Patterns of Flight Operations

Table 13 shows the effective route+ strategic patterns. Three patterns were adopted. The first pattern was adopted by the case NA5. This pattern includes the actions categories of the route, the operating and the fuel. The second pattern was adopted by the case E5. The second pattern includes the actions

Table 13. Effective Route+ Strategic Patterns of Flight Operations

Case ID	Strategy pattern	Action category Total Impact	Significant actions taken	Average Relative % of change in Direct GHG emissions
NA5	R-O-F	**Route (R)** Total Impact= -8.8%	PC-calculated flight path	-4.8%
			Using Global Positioning System(GPS)	-4%
		Operating (O) Total Impact= -4.3%	Revised and improved fuel loading regulations	-4.3%
		Fuel (F) Total Impact= -12.5%	Using biofuels	-12.5%
Total impact of the pattern				-25.5%
E5	R-O-W	**Route (R)** Total Impact= -3.5%	PC-calculated flight path	-3.5%
		Operating (O) Total Impact= -12.7%	Reduced use of auxiliary power unit	-2.2%
			Acceleration altitude	-3.5%
			Reducing use of extended landing light	-3.5%
			Reducing landing time	-3.5%
		Weight (W) Total Impact= -10.1%	Using EFB (Electronic Flight Bag)	-3.5%
			Light meal trolleys	-3.9%
			Light weight cargo containers	-2.7%
Total impact of the pattern				-26.3%
SA1	R-O-M	**Route (R)** Total Impact= -11.7%	Broadcasting technology for communication everywhere	-5.4%
			Monitoring actual flight path and relevant data	-6.3%
		Operating (O) Total Impact= -8.5%	Reduced use of auxiliary power unit	-2.4%
			Altitude optimization	-2.8%
			Reducing landing time	-3.3%
		Maintenance (M) Total Impact= -6.9%	Engine washing	-6.9%
Total impact of the pattern				-27.4%

categories of route, operating, and weight. The third pattern was adopted by the case SA1. The third pattern includes the actions categories of route, operating, and maintenance. Two shared actions categories between the three strategic patterns. These categories were the route and the operating. The second and third strategic patterns took more actions categories than the first pattern. The impact of each action category in reducing the direct GHG emissions of the flight operations varies for the first pattern (which was adopted by

NA5). The most effective action category was fuel by an impact of -12.5%. The second effective action category was route by an impact of -8.8%, and the lowest effective action category was operating by an impact of -4.3%. The impact of actions category in reducing the direct GHG emission for the second strategic pattern (which was adopted by E5) also varies. The most effective action category was operating by an impact of -12.7%, followed by weight category by an impact of -10.1%, and the lowest effective action category was the route by an impact of -3.5%. The impact of actions category in reducing the direct GHG emission for the third strategic pattern (which was adopted by SA1) also varies. The most effective action category was route by an impact of -11.7%, followed by operating category by an impact of -8.5%, and the lowest effective action category was the maintenance by an impact of -6.9%.

The number of actions taken by the first strategy pattern varies. The largest number was related to route action categories by two actions. One action was taken related to each category of the operating and the fuel. The route actions which taken by NA5 were PC calculated flight path, and using global positioning system. The impact of these two actions was -4.8% and -4% respectively. The only operating action taken by NA5 was revised and improved fuel loading regulations by an impact of -4.3%. In addition, the only fuel action taken by NA5 was using the biofuels by an impact of -12.5%.

One action related to route category was taken by E5, which was the same as the action taken by NA5 (PC-calculated flight path). The impact of this action in reducing the direct GHG emissions was -3.5%. Four operating actions were taken by E5. These actions were reduced using of the auxiliary power unit, acceleration altitude, reducing use of the extended landing light, and reducing the landing time. The impact of these actions in reducing the direct GHG emissions was -2.2%, -3.5%, -3.5%, -3.5% respectively. The combination of operating actions were taken by E5 differs from NA5. Three weight actions were taken by E5. These actions were using the electronic flight bag, the light meal trolleys, and the light weight cargo container. The impact of these actions on reducing the direct GHG emissions was -3.5%, -3.9% and -2.7% respectively.

Two route category actions were taken by SA1. These actions were broadcasting technology for communication everywhere, and monitoring of actual fight path and relevance data. The impact of these actions were -5.4% and -6.3% respectively. The route actions taken by SA1 were not shared with E2 and NA5. Also, the impact of this action category was more than other cases (NA5 and E5). Three operating actions were taken by SA1. The first action, which was reduce using of the auxiliary power unit, was shared with E5 and the impact was somehow close -2.4%. The remaining operating actions taken were altitude optimization and reducing the landing time. The impact of these two actions in reducing the direct GHG emissions was -2.8% and -3.3% respectively. One maintenance action was taken, which was engine washing by an impact of -6.9%.

Effective Direct GHG Emissions Strategic Patterns of Ground Operations

Table 14 shows the actions indices and strategy patterns of the ground operations. It can be seen that the levels of realized actions indices were low, moderate and high. For the purpose of this study, only the moderate and the high-level actions indices were recognized as significant actions taken by airlines. It can be seen that two cases (NA3 and E1) took a combination of different significant actions. Other cases took only one action as NA5 and NA6. On the other hand, the significant actions were taken by the reduced emissions and the increased emissions cases. The significant actions taken by the surveyed cases were used as a source to identify the strategic patterns of the cases. According the last column of Table 14 shows these patterns. One strategic pattern was adopted, which was design + pattern.

Table 14. Actions Indices and Strategy Patterns of Ground Operations

Level of performance	Case ID	Vehicle and Engine Design (VD)	Vehicle and Engine Operating (VO)	Vehicle and Engine maintenance management (VM)	Transportation management (VT)	Strategy Pattern	Strategy title
Reduced Emissions cases							
+H	E2	0	0.25 (L)	0	0	------	-----
+M	E3	0	0.25 (L)	0	0	------	------
+L	NA3	1 (H)	0.5 (M)	0	0	VD-VO	Design+
+L	E1	1 (H)	0	1 (H)	0	VD-VM	Design+
Increased emissions cases							
-L	ME1	0	0	0	0	-----	-----
-L	A3	0	0	0	0	-----	-----
-L	NA6	1 (H)	0	0	0	VD	Design+
-M	NA5	1 (H)	0	0	0	VD	Design+

(L): Low **(M):** Moderate **(H):** High

Effective Design⁺ Strategic Patterns of Ground Operations

Table 15 shows the effective design+ strategic patterns. Two patterns were adopted. The first pattern was adopted by the case NA3. This pattern includes the actions categories of design and operating. The second pattern was adopted by the case A1. The second pattern includes the actions categories of the design and maintenance. Design actions category was the shared between the two strategic patterns. The impact of each action category in reducing the direct GHG emissions of the ground operations was the same for the first pattern (which was adopted by NA3). The impact was -5.5% for each category. The impact of action categories in reducing the direct GHG emission for the second strategic pattern (which was adopted by A1) varies. The most effective action category was the maintenance by an impact of -13.2%, followed by the operating category by an impact of -6.6%.

Table 15. Effective Design+ Strategic Patterns of Ground Operations

<table>
<tr><th>Case ID</th><th>Strategy pattern</th><th>Action category
Total Impact</th><th>Significant actions taken</th><th>Average Relative % of change in Direct GHG emissions</th></tr>
<tr><td rowspan="2">NA3</td><td rowspan="2">VD-VO</td><td>Design (VD)
Total Impact= -5.5%</td><td>Using electrical carts or tractors (trucks)</td><td>-5.5%</td></tr>
<tr><td>Operating (VO)
Total Impact= -5.5%</td><td>New forklifts</td><td>-5.5%</td></tr>
<tr><td colspan="4">Total impact of the pattern</td><td>-11%</td></tr>
<tr><td rowspan="3">A1</td><td rowspan="3">VD-VM</td><td>Design (VD)
Total Impact= -6.6%</td><td>Using electrical carts or tractors (trucks)</td><td>-6.6%</td></tr>
<tr><td rowspan="2">Maintenance (VM)
Total Impact= -13.2%</td><td>Washing engine with less water</td><td>-6.6%</td></tr>
<tr><td>Cleaning vehicles and equipment</td><td>-6.6%</td></tr>
<tr><td colspan="4">Total impact of the pattern</td><td>-19.8%</td></tr>
</table>

One action was taken by the first strategy pattern related to each action category. The design action which taken by NA3 was using electrical carts or tractors (trucks)). The impact of this action was -5.5%. The operating action taken by NA3 was using new forklifts by an impact of -5.5%. One action related to design category was taken by A1, which was the same as the action taken by NA3 (using electrical carts or tractors (trucks). The impact of this action in reducing the direct GHG emissions was -6.6%. Two maintenance actions were taken by A1. These actions were washing the engine with less water, and cleaning the vehicles and the equipment. The impact of these actions in reducing the direct GHG emissions was -6.6% and -6.6% respectively.

Summary of Effective Direct GHG Emissions Strategic Patterns

Table 16 shows a summary of the effective direct GHG emissions strategic patterns. It can be seen that the most effective pattern was design+ pattern of the flight operations (D-O-W-F-M) which was adopted by ME3. The impact of this pattern in reducing the direct GHG emissions was -37.7%. This was the only pattern adopted somehow all actions category. So this pattern could be called do all strategy. The next ranked patterns, which had close impact in reducing the direct GHG emissions, were rout+ patterns. The impact was between -25.5% to -27.4%. The most effective route+ pattern was (R-O-M)

Table 16. Summary of Airline Effective Direct GHG Emission Strategic pattern

Case ID	Strategy pattern title	Strategy Pattern	Relative average % of change in Direct GHG emissions	Aircraft/ Engine and vehicle design (D) / (VD)	Flights routes management (R)	Flight/ Engine and vehicle Operating (O) / (VO)	Flights weight management (W)	Flights fuel management (F)	Flights/ Engine and vehicle maintenance management (M) / (VM)
ME3	Design +	D-R-O-W-F-M	-37.7% (H)	-5.2%	-10.5%	-5%	-17.2%		-2.3%
SA1	Route +	R-O-M	-27.4% (H)		-11.7%	-8.5%			-6.9%
E5	Rout +	R-O-W	-26.3% (H)		-3.5%	12.7%	-10.1%		
NA5	Rout +	R-F	-25.5% (H)		-8.8%	-4.3%		-12.5%	
E3	Fuel	F	-25% (M)					-25%	
E1	Design + (ground operations)	VD-VM	-19.8% (M)	-6.6%					-13.2%
A7	Design +	D-R-M	-13.3% (M)	-5.5%	-10.2%				-4.8%
NA3	Design + (ground operations)	VD-VO	-11% (M)	-5.5%		-5.5%			
A5	Operating +	O	-5.4% (L)			-5.4%			
E1	Operating +	O-F	-2.4% (L)			-2%		-0.4%	

(L): Low **(M):** Moderate **(H):** High

Note: to have a better insight about the strategy patterns' acronyms, please see Table 6.

by an impact of -27.4%, followed by route+ pattern (R-O-W) by an impact of -26.3%. In addition, the lowest effective pattern was (R-F) by an impact of -25.5%.

Fuel pattern of flight operations (F) was the fifth ranked by an impact of -25%. The operating+ patterns of flight operations were the lowest rank. The pattern (O) had an impact of -5.4% with the ninth order. However, the pattern (O-F) had an impact of -2.4% with the tenth order. The two ground operations patterns were seventh and eighth ranked. The pattern design+ of the ground operations (VD-VM) was more effective than (VD-VO). The pattern (VD-VM) reduced the direct GHG emissions by -19.8%. However, the pattern (VD-VO) reduced the direct GHG emissions by -11%.

DISCUSSION

About 50% of airlines have an increase in the direct GHG emissions for both the flight and the ground operations. In addition, the majority of reduced emissions airlines achieved a low level of reduction. This result could be justified in the context of demand. The demand is increasing in airlines overtime. According to updated statistics by IATA (International Air Transport Association), the number of international passengers in Jan. 2019 grew by 6% in comparison with Jan. 2018. All regions recorded demand growth, with a capacity increasing by 5.8%. In addition, the domestic market traffic rose 7.3% overall year-on-year (IATA, 2019b).

Different patterns of the flight and the ground green operations strategy were adopted. The adopting of green operations strategy is affected by many operational factors as load factor, length of each flight (Reals, 2007), passenger occupancy, the types of aircraft, the seating capacity, the cargo capacity, the engines and winglets (Atmosfair, 2018). The patterns of the flight operations were more diverse and more effective than the ground operations, since the ground operations are less important than the flight operations. The ground operations account for about 1% or less of the aviation emissions (Vigilante, 2009). In addition, some patterns were shared between the increased and the reduced emissions cases.

The most effective design+ pattern was the pattern adopted more diverse actions. This pattern adopted the design, the operating, the route, the weight and the maintenance action categories. However, the second pattern adopted the design, the route and the maintenance action categories. Both patterns share the same impact of route category, also they adopted the same actions, but there are differences in the adopted design actions. The first pattern adopted the shark let wing tips and the split scimitar. However, the second pattern adopted engine modifications. The impact of both patterns actions are very close and the rank of the action category was third for both. This indicates the differences between design actions is minor and the impact of design action is still not effective as the weight and the route actions.

The relative design actions change is somehow fixed over time, once the engine is replaced or the wings design is changed, the impact will be somehow fixed over time, since the technology itself is fixed. However, other flight actions have different impact over time, since they are flexible actions. The weight action of the first pattern was the most effective category followed by the route management. The weight management is one of the most three important factors ranked after engine performance and aerodynamic, that affecting the aircraft consumption of fuel and emissions (Hollinger, 2016). The route actions were the second ranked by the first pattern. This result was confirmed by many researchers. Once the routes are redesigned, the time spent on flight could be shorter, and the emissions will be minimized (Timperley, 2017).

The maintenance management of both patterns was the last ranked action. However, the impact of maintenance management of the second pattern was better than the first pattern, despite both are sharing the same action (engine washing). This could be related to modifying engines as made by the second pattern. Modifying engines could be more effective since some older engines used sludge chambers in the crankshafts, which were manufactured with hollow crankpins that serve as sludge removers. The sludge chambers require inspection and cleaning at overhaul. If the sludge chambers are formed by means of tubes pressed into the hollow crankpins, make certain they are re-installed correctly to avoid covering the ends of the oil passages. Due to improved oils, sludge chambers are no longer used with modern engines (flight-Mechanic, 2017).

The effective fuel management pattern was affected by one significant action. This action was training on the fuel saving. This action face a lot of concern from many academic institution. For example, IATA Training and Development Institute organizes course in aviation fuel management, in these courses the trainee will learn how to manage fuel from refinery toward the delivery into the aircraft. Also, will learn how to minimize the fuel quantities, and manage the process to have safer sourcing (IATA Training and Development Institute, 2019).

Two operating+ patterns were adopted. The first pattern adopted only operating action category. However, the second pattern adopted operating and fuel management actions. The first pattern was more effective than the second. The two patterns adopted same operating action, which was reduce the using of Auxiliary Power Unit (APU). The impact of this action was more effective by the first pattern than the second pattern. The APU running time between landing and take-off differ per aircraft type and airport. To limit APU emissions, several airports limit the running time of APU's and require aircraft to use fixed electricity connection (gate power) and preconditioned air instead (Dellaret & Hulskotte, 2017).

In addition, the first pattern adopted landing continuous descent operations (CDO). The potential CO2 benefits from optimizing European wide CDOs were estimated to be ten times more than those of the optimizing CCO (climb-out). In addition, the potential saving in Europe is up to 350,000 tones of fuel, which is equivalent to 1.1 million tons of CO2 emissions per year. However, the adopting of 100% CDO is impossible because of weather conditions and capacity constraints (EASA, 2016). But the second pattern adopted other operational actions as reducing the aircraft take-off, launching and taxiing. At an airport, emissions from aircraft idling (at the gate and awaiting take-off), taxiing, taking-off and landing represent a significant source of particles (Herndon et al. 2005). The results of previous studies indicated that using reduced thrust takeoff reduces fuel consumption, nitrogen oxides (NO_X) and black carbon (BC) emissions by 1.0–23.2%, 10.7–47.7%, and 49.0–71.7% respectively, depending on aircraft-engine combinations relative to 100% thrust takeoff (Koudis et al. 2017). The second pattern adopted the bio-fuel action, but the impact of this action was the lowest. The adoption of bio-fuel is still limited because of high cost in production. This is the major challenge since the fuel cost represent 30% of airlines operating cost. However, the potential reduction of GHG emissions as a result of operating bio-jet fuels will be more than the fossil-based jet fuel by 50% to 95% (IRENA, 2017).

Three route+ strategy patterns were adopted. The first pattern adopted route, operating and fuel action categories. The second pattern adopted route, operating and weight action categories. However, the third adopted the route, the operating and the maintenance action categories. Accordingly, two actions categories were shared between all patterns, which they were the route and the operating related actions. The three patterns were very close in the total impact of reducing the direct GHG emissions. This could be justified since the actions categories are highly shared between patterns, also there were many shared taken actions. The most effective action category adopted by the first pattern was fuel management. The action taken was using bio-fuels, as discussed earlier using bio-fuel jets will have more significant impact than fossil-based fuel in reducing the direct GHG emissions (IRENA, 2017). This pattern adopted one operating action, which was revised and improved fuel-loading regulations. This actions could be affected by route actions taken by this pattern as using PC to calculate the flight route, and using GPS. These two actions definitely will affect the reduction in using fuel.

The most effective action category adopted by the second pattern was operating. Four actions were taken in this regard by this pattern as reduce using APU, accelerating altitude, reducing the landing time and reduce the using of landing light. All of these actions as discussed earlier are effective as reported by the previous studies. The most effective action category by the third pattern was route management. The most effective action taken by this pattern was monitoring the actual flight path and the relevant data, since regulatory criteria for minimum runway requirements encompass multiple calculations inclusive of Takeoff Distance (TOD), Takeoff Run (TOR) and Accelerate Stop Distance (ASD). The most limiting of these criteria, based on the aircraft weight and prevailing atmospheric conditions, defines the minimum runway required for takeoff (Skybrary, 2019).

Two ground operations strategy patterns were adopted. These were design+. The first pattern adopted design and operating action categories. The second pattern adopted the design and the maintenance action categories. Two design actions were shared between patterns and its impact was somehow close. This action was using the electrical carts or tractors (trucks). This action is the third ranked action in its impact on airport emissions and has a lot of concern from the operators of aviation sector. This action reduced the emissions of vehicles in airport by 28% (Heathrow Airport, 2018). On the other hand, two maintenance actions were responsible about the most reduction in the direct GHG emissions. These actions were cleaning the engines and the equipment, and washing the engines with less water. Washing engine in general will reduce the emissions since the dust will be removed and the engine will work more efficiency. However, washing with water could cause the damage for sensors or can move dirt and cuck toward the belts, which influence the engine effectiveness and cause more emissions.

CONCLUSION

The aim of this chapter was to report the effective airlines green operations strategy in reducing the direct GHG emissions. This research issue has raised by limited studies. Most of the previous studies have reported general green strategy patterns of airlines. The main results of this study were 50% of airlines have an increase in the direct GHG emissions for both the flight and the ground operations. In addition, the majority of reduced emissions by airlines achieved a low level of reduction. Different patterns of the flight and the ground green operations strategy were adopted. The most effective design+ pattern was the pattern adopted more diverse actions. This pattern adopted the design, the operating, the route weight and the maintenance action categories. The second pattern adopted the design, the route and the maintenance action categories. The effective fuel management pattern was affected by one significant action, which was training on the fuel saving.

Two operating+ patterns were adopted. The first pattern adopted only operating actions category. However, the second pattern adopted the operating and the fuel management actions. The first pattern was more effective than the second was. Three route+ strategy pattern were adopted. The first pattern adopted the route, the operating and the fuel action categories. However, the second pattern adopted the route, the operating and the weight action categories. On the other hand, the third adopted the route, the operating and the maintenance action categories. The three rout+ patterns were very close in the total impact on reducing the direct GHG emissions. Two ground operations strategy patterns were adopted. These were design+, the first pattern adopted the design and the operating category. The second pattern adopted the design and the maintenance action categories.

The results of this chapter will benefit the academics and the industry experts alike. The academic have in-depth reporting for the relationship between the actions and the green operations actions. Therefore, they can develop hypotheses for the future researches. The experts in the industry have a clear idea about the different strategy patterns that could be adopted. In addition, the required actions and their impact on reducing the direct GHG emissions. Despite the significant contribution of this chapter, some recommendations could be given for the future researchers. It is recommended to investigate a larger sample size to have more valid results. It is recommended to investigate the interrelationship between the actions categories and the actions taken to have a better understanding for the independent, the moderating, the mediating and the dependent variables.

REFERENCES

Abdullah, M. A., Chew, B. C., & Hamid, S. R. (2016). Benchmarking key success factors for the future green airline industry. *Procedia: Social and Behavioral Sciences*, *224*, 246–253. doi:10.1016/j.sbspro.2016.05.456

Air Transport Action Group. (2018). *Facts and Figures*. https://www.atag.org/facts-figures.html

Amizadeh, F., Alonso, G., Benito, A., & Morales-Alonso, G. (2016). Analysis of the recent evolution of commercial air traffic CO2 emissions and fleet utilization in the six largest national markets of the European Union. *Journal of Air Transport Management*, *55*, 9–19. doi:10.1016/j.jairtraman.2016.04.006

Atmosfair. (2018). *Atmosfair airline index 2018*. https://www.atmosfair.de/wp-content/uploads/aai2018-englischfarbe_final_mn.pdf

Bofinger, H., & Strand, J. (2013). *Calculating the carbon footprint from different classes of air travel*. The World Bank. doi:10.1596/1813-9450-6471

Brueckner, J. K., & Abreu, C. (2017). Airline fuel usage and carbon emissions: Determining factors. *Journal of Air Transport Management*, *62*, 10–17. doi:10.1016/j.jairtraman.2017.01.004

Chapman, M. (2016). Sustaining Reductions in Aircraft Emissions for Canada's Major Airlines. In *Managing in a VUCA World* (pp. 175–193). Cham: Springer. doi:10.1007/978-3-319-16889-0_12

Dellaret, S., & Hulskotte, J. (2017). *Emission of air pollutants from civil aviation in Netherland*. TNO. http://www.emissieregistratie.nl/ erpubliek/documenten/Lucht%20(Air)/ Verkeer%20en%20 Vervoer%20(Transport)/Overig%20 Verkeer%20en%20 Vervoer/Dellaert%20 &%20Hulskotte%20-%202017%20-%20 Emissions%20of%20 air%20pollutants%20 from%20 civil%20avi....pdf

EASA. (2016). *European aviation environmental report 2016*. European Airline Safety Association. https://www.eurocontrol.int/sites/default/files/2019-06/eaer-2019_0.pdf

El-Takriti, S., Pavlenko, N., & Searle, S. (2017). *Mitigating international aviation emissions: Risks and opportunities for alternative jet fuels.* White paper, International Council on Clean Transportation.

Flight-Mechanic. (2017). *Engine Cleaning in engine maintenance and operating.* https://www.flight-mechanic.com/engine-cleaning/

Fuglestvedt, J., Berntsen, T., Myhre, G., Rypdal, K., & Skeie, R. B. (2008). Climate forcing from the transport sectors. *Proceedings of the National Academy of Sciences of the United States of America, 105*(2), 454–458. doi:10.1073/pnas.0702958104 PMID:18180450

Givoni, M., & Rietveld, P. (2010). The environmental implications of airlines' choice of aircraft size. *Journal of Air Transport Management, 16*(3), 159–167. doi:10.1016/j.jairtraman.2009.07.010

GRI. (2017). *About GRI.* https:// www.globalreporting.org/Information/about-gri/Pages/default.aspx

Heathrow Airport. (2018). *Emissions strategy and action plan.* https://www.heathrow.com/file_source/Company/Static/PDF/Communityandenvironment/Heathrow_Emissions_Strategy.pdf

Herndon, S. C., Onasch, T. B., Frank, B. P., Marr, L. C., Jayne, J. T., Canagaratna, M. R., ... Miake-Lye, R. C. (2005). Particulate emissions from in-use commercial aircraft. *Aerosol Science and Technology, 39*(8), 799–809. doi:10.1080/02786820500247363

Hollinger, P. (2016*). Airlines bid to beat their weight problem.* https://www.ft.com/content/6ce66d16-bd6a-11e6-8b45-b8b81dd5d080

IATA. (2019a). *Improving environmental performance.* https://www.iata.org/whatwedo/environment/Pages/index.aspx

IATA. (2019b). *Passenger demand begins positively in 2019.* https://airlines.iata.org/news/passenger-demand-begins-positively-in-2019%C2%A0

IATA Training and Development Institute. (2019). *Aviation fuel management essentials.* https://ciiasa.asa.gob.mx/cursos/va/training-talf06-aviation-fuel-management.pdf

IRENA. (2017). *Biofuels for Aviation: Technology Brief.* https://www.irena.org/documentdownloads/publications/irena_biofuels_for_aviation_2017.pdf

Jardine, C. N. (2009). *Calculating the carbon dioxide emissions of flights.* Final report by the Environmental Change Institute.

Kollmuss, A., & Crimmins, A. M. (2009). *Carbon Offsetting & Air Travel Part 2: Non-CO2 Emissions Calculations.* Stockholm Environment Institute, Discussion Paper.

Koudis, G. S., Hu, S. J., Majumdar, A., Jones, R., & Stettler, M. E. (2017). Airport emissions reductions from reduced thrust takeoff operations. *Transportation Research Part D, Transport and Environment*, *52*, 15–28. doi:10.1016/j.trd.2017.02.004

Lee, D. S., Fahey, D. W., Forster, P. M., Newton, P. J., Wit, R. C., Lim, L. L., ... Sausen, R. (2009). Aviation and global climate change in the 21st century. *Atmospheric Environment*, *43*(22-23), 3520–3537. doi:10.1016/j.atmosenv.2009.04.024

Lee, K. C., Tsai, W. H., Yang, C. H., & Lin, Y. Z. (2018). An MCDM approach for selecting green aviation fleet program management strategies under multi-resource limitations. *Journal of Air Transport Management*, *68*, 76–85. doi:10.1016/j.jairtraman.2017.06.011

Liu, X., Zhou, D., Zhou, P., & Wang, Q. (2017). Dynamic carbon emission performance of Chinese airlines: A global Malmquist index analysis. *Journal of Air Transport Management*, *65*, 99–109. doi:10.1016/j.jairtraman.2017.09.009

McCollum, D. L., Gould, G., & Greene, D. L. (2010). *Greenhouse gas emissions from aviation and marine transportation: Mitigation potential and policies.* White paper, Pew Centre on Global Climate Change.

Migdadi, Y. K. A. A. (2018). Identifying the best practices of airlines' green operations strategy: A cross-regional worldwide survey. *Environmental Quality Management*, *28*(1), 21–32. doi:10.1002/tqem.21575

Migdadi, Y. K. A. A. (2020). Green Operational Strategy for Airlines: Content and Regional Analysis. In Cases on Green Energy and Sustainable Development (pp. 193-229). IGI Global.

Reals, K. (2007). *Seen to be green: how airlines are building the environment into their core strategy.* https://www.flightglobal.com/news/articles/seen-to-be-green-how-airlines-are-building-the-envi-220281/

Skybrary. (2019). *Engine Failure during Takeoff - Multi-Engine Transport Category Jet Aircraft*. https://www.skybrary.aero/index.php/Engine_Failure_During_Takeoff_-_Multi-Engine_Transport_Category_Jet_Aircraft

Timperley, J. (2017). *Explainer: The challenges of tackling aviation's non-CO2 emissions*. https://www.carbonbrief.org/explainer-challenge-tackling-aviations-non-co2-emissions

UN-Secretary-General-High-Level Advisory Group on Sustainable Transportation. (2015). *Mobilizing sustainable transport for development*. https://sustainabledevelopment.un.org/content/documents/2375Mobilizing%20Sustainable%20Transport.pdf

Vigilante, M. (2009). *Reducing the Carbon Footprint—What Works*. Transportation Research Board Annual Meeting, Washington, DC.

Yan, W., Cui, Z., & Gil, M. J. Á. (2016). Assessing the impact of environmental innovation in the airline industry: An empirical study of emerging market economies. *Environmental Innovation and Societal Transitions*, *21*, 80–94. doi:10.1016/j.eist.2016.04.001

APPENDIX

How to Assess the Direct GHG Emissions Performance

Table 17. Example on How to Assess the Direct GHG Emissions Performance

Case ID	Phase 1: % of change in emissions	Phase 2: The relative performance	Phase 2: 5-points rating of change in emissions	Phase 3: Assigning rating level
	[(P 2016- P2013)/ P2013]X100%	% of change of any case/ % of change of the best	Relative performance X 5	Rating level
E3	-21%	-21%/-21%= 1.00	5.0	+H
ME3	-20%	-20%/-21%= 0.95	4.7	+H
E6	-7%	-7%/-21%=0.33	1.6	+L
A5	-5%	-5%/-21%= 0.24	1.2	+L
E5	-4%	-4%/-21%= 0.19	0.9	+L
NA5	-2%	-2%/-21%= 0.10	0.6	+L
E1	-2%	-2%/-21%=0.10	0.5	+L
SA1	-2%	-2%/-21%=0.10	0.5	+L
A7	-2%	-2%/-21%= 0.10	0.4	+L
NA2	-2%	-2%/-21%= 0.10	0.4	+L
A6	+0.3%	+0.3/-21%= -0.01	-0.1	-L
NA1	+4%	+4%/-21%= -0.19	-0.6	-L
A1	+4%	+4%/-21%= -0.19	-0.6	-L
E2	+5%	+5%/-21%= -0.24	-0.7	-L
E4	+5%	+5%/-21%= -0.24	-0.8	-L
NA3	+9%	+9%/-21%= -0.43	-1.4	-L
A3	+9%	+9%/-21%= -0.43	-1.4	-L
NA6	+13%	+13/-21%= -0.62	-1.9	-M
NA4	+16%	+16%/-21%= -0.76	-2.4	-M
ME2	+20%	+20%/-21%= -0.95	-3.0	-M
ME1	+34%	+34%/-21%= -1.00	-5.0	-H
Best Performance	-21%			

How to Identify the Effective Direct GHG Emissions Strategic Patterns

Table 18. Example on How to Compute Action Index and Transform it into 5-points scale

Actions of Flight Routes	**NA1**	**NA3**	**NA4**	**NA5**	**NA6**	**E3**	**E5**	**A1**	**A2**	**A3**	**A4**	**ME1**	**A6**	**A7**	**ME3**	**SA1**	
Routes optimization	1	1				1	1	1	1	1	1	1	1	1	1	1	
PC calculate flight path			1	1			1		1		1				1		
broadcasting technology for communicate everywhere			1												1	1	
monitoring actual flight path and relevant data		1												1	1	1	
real time data text communication			1														
using GPS				1	1							1					
center of gravity										1	1					1	
Number of actions taken	1	2	3	2	1	1	2	1	2	2	3	2	1	2	4	4	Best index value
Phase 1: Action index	1/7 0.14	2/7 0.29	3/7 0.43	2/7 0.29	1/7 0.14	1/7 0.14	2/7 0.29	1/7 0.14	2/7 0.29	2/7 0.29	3/7 0.43	2/7 0.29	1/7 0.14	2/7 0.29	4/7 0.57	4/7 0.57	0.57
Relative index value	0.14/0.57 0.25	0.29/0.57 0.51	0.43/0.57 0.75	0.29/0.57 0.51	0.14/0.57 0.25	0.14/0.57 0.25	0.29/0.57 0.51	0.14/0.57 0.25	0.29/0.57 0.51	0.29/0.57 0.51	0.43/0.57 0.75	0.29/0.57 0.51	0.14/0.57 0.25	0.29/0.57 0.51	0.57/0.57 1.00	0.57/0.57 1.00	
Phase 2: 5-points index value	1.25	2.5	3.75	2.5	1.25	1.25	2.5	1.25	2.5	2.5	3.75	2.5	1.25	2.5	5	5	
Phase 3: Assign rating level	L	M	H	M	M	L	M	L	M	M	H	M	L	M	H	H	

How to Compute the Impact of Actions Taken by the Effective Patterns on the Achieved Direct GHG Emissions

Table 19. Example on How to Identify Actions with Moderate and High Indices for Each Case

Rating level	Case ID	Aircraft design (D)	Flights routes management (R)	Flight Operating (O)	Flights weight management (W)	Flights fuel management (F)	Flights maintenance management (M)	Phase 5: Identifying the moderate to high actions indices
Reduced Emissions Cases								
+H	E3	1.7	1.23	1.13	1.39	5	0	F
+H	ME3	5.0	5.00	3.51	5.00	5	2.5	D-R-O-W-M
+L	E6	0.0	0.00	1.38	0.74	0	2.5	M
+L	A5	0.0	0.00	2.87	0.00	0	2.5	O-M
+L	E5	1.67	2.54	2.87	2.13	0	0	R-O-W
+L	NA5	1.25	2.54	2.13	0.74	2.5	0	R-O-F
+L	E1	1.67	0.00	4.26	1.39	2.5	2.5	O-F-M
+L	SA1	1.67	5.00	2.87	1.39	0	2.5	R-O-M
+L	A7	3.3	2.54	0.74	0.00	0	2.5	D-R-M
+L	NA2	1.67	0.00	0.00	0.00	0	0	
Increased Emissions Cases								
-L	A6	1.67	1.23	2.87	2.78	5	5	O-W-F-M
-L	NA1	1.67	1.23	0.74	1.39	2.5	0	F
-L	A1	3.3	1.23	0.74	1.39	0	0	D
-L	E2	1.67	0.00	1.38	0.74	0	0	
-L	E4	0.0	0.00	1.38	0.00	0	0	
-L	NA3	0.0	2.19	0.74	2.13	0	0	R-W
-L	A3	0.0	2.54	1.38	1.39	0	2.5	R-M
-M	NA6	1.67	3.77	5.00	2.13	0	5	R-O-W-M
-M	NA4	3.3	1.23	0.00	2.87	2.5	2.5	D-W-F-M
-M	ME2	3.3	3.77	0.74	0.00	0	0	D-R
-H	ME1	3.3	0.00	1.38	1.39	2.5	0	D-F

Table 20. Example on Phase 6 & 7: How to Identify Cases of Design+ Patterns

Rating level	Case ID	Aircraft design (D)	Flights routes management (R)	Flight Operating (O)	Flights weight management (W)	Flights fuel management (F)	Flights maintenance management (M)	Patterns
Reduced Emissions Cases								
+H	ME3	5.0	5.00	3.51	5.00	5	2.5	D-R-O-W-M
+L	A7	3.3	2.54	0.74	0.00	0	2.5	D-R-M
Increased Emissions Cases								
-L	A1	3.3	1.23	0.74	1.39	0	0	D
-M	NA4	3.3	1.23	0.00	2.87	2.5	2.5	D-W-F-M
-M	ME2	3.3	3.77	0.74	0.00	0	0	D-R
-H	ME1	3.3	0.00	1.38	1.39	2.5	0	D-F
Number of shared actions		5 100%	2 33%	1 15%	2 33%	3 50%	3 50%	Pattern title: Design+

Table 21. Example on Phase 6 & 7: How to Identify the Cases of Fuel Patterns

Rating level	Case ID	Aircraft design (D)	Flights routes management (R)	Flight Operating (O)	Flights weight management (W)	Flights fuel management (F)	Flights maintenance management (M)	Patterns
Reduced Emissions Cases								
+H	E3	1.7	1.23	1.13	1.39	5	0	F
Increased Emissions Cases								
-L	NA1	1.67	1.23	0.74	1.39	2.5	0	F
Number of shared actions						2 100%		Pattern title: Fuel

Table 22. Example on Phase 6 & 7: How to Identify the Cases of Operating+ Patterns

Rating level	Case ID	Aircraft design (D)	Flights routes management (R)	Flight Operating (O)	Flights weight management (W)	Flights fuel management (F)	Flights maintenance management (M)	Patterns
Reduced Emissions Cases								
+L	A5	0.0	0.00	2.87	0.00	0	2.5	O-M
+L	E1	1.67	0.00	4.26	1.39	2.5	2.5	O-F-M
Increased Emissions Cases								
-L	A6	1.67	1.23	2.87	2.78	5	5	O-W-F-M
Number of shared actions				3 100%	1 33%	2 66%	3 100%	Pattern title: Operating+

Table 23. Example on Phase 6 & 7: How to Identify the cases of Maintenance Pattern

Rating level	Case ID	Aircraft design (D)	Flights routes management (R)	Flight Operating (O)	Flights weight management (W)	Flights fuel management (F)	Flights maintenance management (M)	Patterns
Reduced Emissions Cases								
+L	E6	0.0	0.00	1.38	0.74	0	2.5	M
Number of shared actions							1 100%	Pattern title: Maintenance

Table 24. Example on Phase 6 & 7: How to Identify the Cases of Route+ Patterns

Rating level	Case ID	Aircraft design (D)	Flights routes management (R)	Flight Operating (O)	Flights weight management (W)	Flights fuel management (F)	Flights maintenance management (M)	Patterns
Reduced Emissions Cases								
+L	E5	1.67	2.54	2.87	**2.13**	0	0	R-O-W
+L	NA5	1.25	2.54	2.13	0.74	2.5	0	R-O-F
Increased Emissions Cases								
-L	NA3	0.0	2.19	0.74	2.13	0	0	R-W
-L	A3	0.0	2.54	1.38	1.39	0	2.5	R-M
-M	NA6	1.67	3.77	5.00	2.13	0	5	R-O-W-M
Number of shared actions			5 100%	3 60%	3 60%		2 40%	Pattern title: Route+

Table 25. Example on Phase 1: How to Identify the Significant Actions taken by the Effective Case

Action category	Effective case	Actions made by the effective case	Benchmarked lower effective or ineffective cases				
			A7	A1	NA4	NA6	ME2
Design	ME 3	Shark let wing tips	Taken	Taken		Taken	Taken
	ME 3	Split scimitar winglets			TAKEN		
Route	ME 3	PC-calculated flight path	Taken	Taken		Taken	Taken
	ME 3	Broadcasting technology for communication everywhere	Taken	Taken		Taken	Taken
	ME 3	Route optimization			TAKEN	TAKEN	TAKEN
	ME 3	Monitoring actual flight path and relevant data		TAKEN	TAKEN	TAKEN	TAKEN
Operating	ME 3	Acceleration altitude	Taken	Taken	Taken	Taken	Taken
	ME 3	Accurate planning of flight	Taken	Taken	Taken	Taken	Taken
Weight	ME 3	Optimal water potable carried	Taken	Taken	Taken		
	ME 3	Using EFB (Electronic Flight Bag)	Taken	Taken	Taken	Taken	Taken
	ME 3	Using onboard tablets	Taken	Taken	Taken	Taken	Taken
	ME 3	Light meal trolleys	Taken		Taken		Taken
	ME 3	Eliminating magazines	Taken	Taken	Taken	Taken	
	ME 3	Light weight cargo containers	Taken	Taken	Taken		Taken
	ME 3	Restricting fuel overload	Taken	Taken	Taken		Taken
Maintenance	ME 3	Engine washing		TAKEN	TAKEN		TAKEN

Table 27. Example on Phase 3: How to Compute the Relative Impact of Actions Categories

Action category	Effective case	Benchmarked lower effective or ineffective cases				
		A7	A1	NA4	NA6	ME2
Design	ME 3	1/12= 0.08	1/13= 0.08	1/13= 0.08	1/10= 0.10	1/13= 0.08
Route	ME 3	2/12= 0.17	3/13= 0.23	2/13= 0.15	4/10= 0.40	4/13= 0.31
Operating	ME 3	2/12= 0.17	2/13= 0.15	2/13= 0.15	2/10= 0.20	2/13= 0.15
Weight	ME 3	7/12= 0.58	6/13= 0.46	7/13= 0.54	3/10= 0.30	5/13= 0.38
Maintenance	ME 3		1/13= 0.08	1/13= 0.08		1/13= 0.08

Table 26. Example on Phase 2: How to Count the Number of Actions Taken by the Effective Case

Action category	Effective case	Benchmarked lower effective or ineffective cases				
		A7	A1	NA4	NA6	ME2
Design	ME 3	1	1	1	1	1
Route	ME 3	2	3	2	4	4
Operating	ME 3	2	2	2	2	2
Weight	ME 3	7	6	7	3	5
Maintenance	ME 3		1	1		1
Total		12	13	13	10	13

Table 28. Example on Phase 4: How to Compute the Differences in Direct GHG Emissions of the Effective Case in Comparison with Ineffective Cases

Effective case	Benchmarked lower effective or ineffective cases				
	A7	A1	NA4	NA6	ME2
ME 3	-18%	-24%	-36%	-33%	-40%

Table 29. Example on Phase 5: How to Compute the Reduction in Direct GHG Emissions for Each Action Category of the Effective Case

Action category	Effective case	Benchmarked lower effective or ineffective cases				
		A7	A1	NA4	NA6	ME2
Design	ME 3	0.08X-18%= -1.5%	0.08X-25%= -2%	-3%	-3%	-3%
Route	ME 3	0.17X18%= -3%	0.23X-24%= -6%	-6%	-13%	-12%
Operating	ME 3	0.17X-18%= -3%	0.15X-24%= -4%	-6%	-7%	-6%
Weight	ME 3	0.58X-18% -10.5%	0.46X-24%= -11%	-19%	-10%	-15%
Maintenance	ME 3		0.08X-24%= -2%	-3%		-3%
Total		-18%	-24%	-36%	-33%	-40%

Table 30. Example on Phase 6 & 7: How to Compute the Reduction in Direct GHG Emissions for Each Significant Action Taken by the Effective Case

Action category	Actions made by the effective case	A7	A1	NA4	NA6	ME2	Phase 7: Average
Design	Shark let wing tips	-1.5%	-1.85%		-3.30%	-3.08%	-2.4%
	Split scimitar winglets			-2.77%			-2.8%
Route	PC-calculated flight path	-3%/2= -1.5%	-1.85%		-3.30%	-3.08%	-2.4%
	Broadcasting technology for communication everywhere	-3%/2= -1.5%	-1.85%		-3.30%	-3.08%	-2.4%
	Route optimization			-2.77%	-3.30%	-3.08%	-3%
	Monitoring actual flight path and relevant data		-1.85%	-2.77%	-3.30%	-3.08%	-2.7%
Operating	Acceleration altitude	-3%/2= -1.5%	-1.85%	-2.77%	-3.30%	-3.08%	-2.5%
	Accurate planning of flight	-3%/2= 1.5%	-1.85%	-2.77%	-3.30%	-3.08%	-2.5%
Weight	Optimal water potable carried	-11%/7= -1.5%	-1.85%	-2.77%			-2%
	Using EFB (Electronic Flight Bag)	-11%/7= -1.5%	-1.85%	-2.77%	-3.30%	-3.08%	-2.5%
	Using onboard tablets	-11%/7= 1.5%	-1.85%	-2.77%	-3.30%	-3.08%	-2.5%
	Light meal trolleys	-11%/7= 1.5%	-1.85%	-2.77%		-3.08%	-2.3%
	Eliminating magazines	-11%/7= 1.5%	-1.85%	-2.77%	-3.30%		-2.4%
	Light weight cargo containers	-11%/7= 1.5%		-2.77%		-3.08%	-2.5%
	Restricting fuel overload	-11%/7= -1.5%	-1.85%	-2.77%		-3.08%	-3%
Maintenance	Engine washing	-1.5%	-1.85%	-2.77%		-3.08%	-2.3%

Chapter 3
Airline Effective Indirect Greenhouse Gases Emissions Strategies

ABSTRACT

This chapter reports the effective indirect GHG2 emissions strategies. To achieve this aim, the practices of 18 airlines from five regions were investigated. The data used in this chapter was the secondary published data in the annual sustainability reports. The data was retrieved from GRI (Global Reporting Initiatives) website. The special tailored data analysis techniques were used for this purpose. This chapter reported eight effective strategic patterns. The strategic patterns were four energy saving+ patterns, two sustainable energy+ patterns, one replacing facilities pattern, and one accreditation+ pattern. This chapter helps decision makers and academics alike, since the impact of effective actions on reducing the indirect GHG2 emissions of each pattern were reported.

INTRODUCTION

The energy conservation output and the emissions reduction of airports buildings are not sufficient and should be expanded in the future (Jun & Jie, 2018). The percentage of reduction in the energy used by airlines is on average 1.9% and the reduction in the indirect emissions was on average -0.7% (Migdadi, 2018). Accordingly, the aviation industry will be more

DOI: 10.4018/978-1-7998-4255-2.ch003

responsible about making more actions to reduce the negative environmental impact. All stakeholders must make a genuine commitment to transform the transport system (UN-Secretary-General-High-Level Advisory Group on Sustainable Transportation, 2015). In this regard, the airlines should take more effective actions to reduce the indirect GHG2 emissions. The actions taken by airlines are diverse and numerous, but the focus should be on the most effective actions to be adopted.

The most of previous studies have investigated the green actions that will reduce the indirect GHG2 emissions in the airports (e.g. Ortega Alba & Manana, 2016; Sukumaran & Suhakar, 2017; Baxter, Srisaeng & Wild, 2018a,b; Xu & Deng, 2018; Baxter, Srisaeng & Wild, 2019). The limited studies that investigated the indirect GHG2 emissions have reported the general strategy patterns adopted by the airlines. So, reporting the effective practices of indirect GHG2 emissions was a part of the general scope of some studies (e.g. Migdadi, 2018, 2019 & 2020). This chapter bridged the gap of previous studies by reporting in-depth the effective green operations strategies in reducing the indirect GHG2 emissions. This chapter reported the practices of many airlines in different regions and developed strategic patterns. This chapter adopted the benchmarking methodology in making comparison and reporting the most effective patterns. Accordingly, the main question of this chapter was: what were the most effective green strategic patterns adopted by airlines to reduce the indirect GHG2 emissions? To answer the question the following objectives were realized:

1. Ranking the airlines according to its indirect GHG2 emissions performance.
2. Identifying the effective indirect GHG2 emissions strategic patterns that adopted by airlines.
3. Identifying the significant green actions taken by the effective strategic patterns.
4. Identifying the impact of significant green actions on reducing the indirect GHG2 emissions.

This chapter is arranged in five sections: the first one is the literature review, followed by the methodology and methods, then, the results of the empirical study, next, the discussion, and the final section is the conclusions.

LITERATURE REVIEW

This part of the chapter discussed the key literature of the airlines indirect GHG2 emissions. The first part of the literature aimed to discuss how the indirect GHG2 emissions are measured. So, different methodologies and models were presented. Also, the factors and assumptions that should be considered in measuring the indirect GHG2 emissions were discussed. The second part of the literature review discussed the green operations actions lists, categories, and the strategy patterns.

AIRLINE INDIRECT GHG2 EMISSIONS

GHG emissions as discussed in the previous chapter could be classified into direct and indirect emissions. The direct emissions are the emissions from the jet fuels and the ground support activities. However, the indirect GHG emissions are a consequence of the particular company (airline) activities, but occur at sources owned or controlled by another company. The indirect emissions could be classified into GHG2 and GHG3 emissions. GHG2 emissions, or what is called scope 2 emissions, are the emissions from the purchased and consumed heat, electricity or steam (Sotos, 2015). It is the metric tons of indirect CO2 emissions of electricity, power and heat from the direct billing of owned or leased facilities (Migdadi, 2018, 2019, 2020).

The assessor of emissions should take care during the reporting of such kind of emissions, because sometimes the situation could be ambiguous, and the repotting could be duplicated or incorrect. For example, some companies such as the electricity utilities or suppliers may own the energy generation facilities that sell all their power through the local grid. The emissions from the generation facilities are reported under the scope 1. However, the same company could have separated administrative and commercial or industrial facilities or offices buildings that consume electricity from the same grid, in this case the emissions of these facilities will be reported under the scope 2 (Sotos, 2015).

The definition of GHG emissions requires more attention while studying the airlines emissions. For example, the facilities that are managed in the hub airport and are supplied of the electricity indirectly through the local electrical grid, in this case, the facilities and the buildings emissions could be attributed as the scope 2 emissions. However, if the power is generated by the generators located in the airports, in this case, the emissions of facilities and buildings could be attributed as the scope 1 emissions. On the other hand, at the spokes airport, the situation is more complicated and requires more in-depth investigation to check the sources of power and heat.

The facilities as the buildings, the offices, the desks, the vehicles, and the engines at spokes airports are almost leased. Accordingly, the supply of electricity power and heat will be indirect through paying the fees for the leased services at airports. In addition, the assessor needs to be more accurate in the investigation and check if these leased services are under the full control of the airline itself, if so, the scope of emissions will be the scope 2. However, if the airline is not operating these facilities, the scope will be the scope 3. The papers of Migdadi (2018, 2019, 2020) assumes that the facilities, the buildings, the vehicles and the engines are located in the spokes airports, so they are leased and under the control of airlines.

The literature of airport energy management discussed this issue more than the literature of airlines green operations literature. Accordingly, the sources of energy of airports are the electricity and the fuel used for heating the boiler or the vehicles. Electricity is the main source of energy, because it is necessary for the supply of main airport energy consumers. Electricity is almost comes from the commercial grid and is supplied by a power company (Ortega Alba & Manana, 2016). The energy consumption at airports is affected by the season of the year, also differs over the day hours. The highest consumption of electricity is between 6:00 am. to 9:00 am. and between 7:00 pm. to 9:30 pm. (Ortega Alba and Manana, 2017). Moreover, the building characteristics play a dynamic role in the energy consumption, for example, the building shape, the compactness, the transparent, the building materials etc. play such role in the energy consumption (Ortega Alba & Manana, 2016).

In the airports, the terminal buildings are the highest energy consumption. The most intensive facilities for the energy consumptions are HVAC (Heat, Ventilation, and Air Conditioning) system and the lighting systems (Ortega Alba & Manana, 2017). HAVC consumes about 24.5% of the energy. However, the lighting consumes about 19.82% of the energy (Ortega Alba & Manana, 2016). According to Migdadi (2018), the percentage of reduction in the indirect GHG2 emissions by airlines was low by about -0.7%. However, the significant cluster in reducing the indirect GHG2 emissions achieved -12% reduction in the emissions. On the other hand, according to more in-depth study by Migdadi (2019), two strategic patterns adopted a significant decrease in the indirect GH2 emissions. The percentage of change was moderate -13% by one strategic pattern (realized by six airlines), and another pattern realized low reduction by -5% (realized by two cases). Accordingly, the average reduction of the indirect GHG2 emissions by airlines sector was -0.7%, but the significant reduction as adopted by the effective cases was between -5% to -13%. In this chapter, the definition of Migdadi (2018, 2019 & 2020) was adopted, but it is restricted for the facilities and the buildings. The vehicles and the engines operations are recognized as a part of the ground support operations at the hub airport, accordingly, its emissions are assumed as direct GHG1 emissions.

Airlines Indirect GHG2 Emissions Operational Actions and Strategy Patterns

The operational actions of indirect GHG2 emissions could be classified into the facilities and the buildings energy saving, the facilities and the buildings accreditation, the facilities and the buildings sustainable energy, upgrading and replacing the facilities (Migdadi, 2018, 2019 & 2020). The previous classification includes also the vehicles and the engines actions as the vehicles and the engine design, the operating, the maintenance and the transportation management. In this study the last actions categories were excluded since the assumption of this chapter is the ground and the support services are operated by the airlines themselves at the hub-airports.

The most of previous studies were investigated the practices of energy consumptions at the airports. The emphasis was on the impact of adopting the sustainable energy (e.g. Ortega Alba & Manana, 2016; Sukumaran & Suhakar, 2017; Baxter, Srisaeng & Wild, 2018a,b; Xu & Deng, 2018; Baxter, Srisaeng & Wild, 2019). According to these studies, the sustainable sources of energy that could be adopted by the airports are numerous such as the using of solar panels (e.g. Budd, budd, & Ison, 2015), the using of wind energy (e.g. Budd, Budd & Ison, 2015), the using of hydro electrical (water) (e.g. Baxter & Srisaeng, 2018b), the using of cogeneration plants (e.g. Baxter, Srisaeng & Wild 2018b), and the using of geothermal. The adopting of solar photovoltaic (PV) systems could reduce the energy consumption by 10% and the emissions by 915 tons (Baxter, Srisaeng & Wild, 2019).

The effectiveness of solar power depends on the power capacity of the system and the size of the airport buildings. For example, the adopting of 12 MWP photovoltaic power is able to generate 12600 MWH and reduce the GHG emissions by 12,569 tons. This was a proposed solar panel for Beijing new airport (Xu & Deng, 2018). Adopting 2MWP plant is capable to generate 2733.122 MWH of the energy a year. This was a proposed solar panel for Raja, Bhoj airport, this system is capable to reduce the emissions by 59200 tons (Sukumaran & Suhakar, 2017). Other sources of the sustainable energy could be adopted. For example, the adoption of cogeneration plans improved the efficiency of energy by 85% (Ortega Alba & Manana, 2016). The using of wind power generators or construct a commercial scale wind turbines (Budd, Budd & Ison, 2015), could generate 1.191 KW a year (Baxter, Srisaeng & Wild, 2018b).

Other significant actions taken by the airports to reduce the energy consumptions were the using LEDs lights (Budd, Budd & Ison, 2015; Malik, 2017; Jun & Jie, 2018). These lights are expected to reduce the energy consumption by 70-90% (Baxter, Srisaeng & Wild, 2018b). it is recommended to be replaced everywhere, in the terminal (inside the buildings) and the airfield (outside the buildings) (Budd, Budd & Ison, 2015; Baxter, Srisaeng & Wild, 2018a). The most significant savings in the airport energy consumptions come from the airport ventilations and cooling (including ATES) (Baxter, Srisaeng & Wild, 2018b). Accordingly, it is recommended to optimize the energy consumption of ventilation system and the cooling towers (Malik, 2017). For example, using the automated energy-saving air conditioning, and linking on-off function of AC equipment with the airport passengers IS, this action reduced 1500 tons of the emissions a year (Baxter, Srisaeng & Wild, 2018b). In addition, it is recommended to renovate the heating/cooling modifications and the pipe network (Jun & Jie, 2018). Replacing the existing boilers with the energy efficient ones (Budd, Budd & Ison, 2015).

Design and upgrade the buildings could reduce the energy consumptions (Ferrulli, 2015). For example, using new materials or technology for the buildings to save energy (Jun & Jie, 2018). Monitoring of the energy consumption (Baxter, Srisaens & Wind, 2018b) as installing of the real time metering system could help in this (Budd, Budd & Ison, 2015). Moreover, setting the energy saving targets and the voltage regulations are proposed actions to reduce the energy consumptions (Baxter, Srisaens & Wild, 2018b). In order to have the comprehensive sustainable buildings design, the airports could adopt LEED standards. These standards have been advised by US GBC (United States Green Building Council) since 1988. The standards have an accreditation system to insure that the buildings are ecologically oriented to provide a healthy work environment. About 270 buildings belong to airports are LEED certified or in process of being certified, about 31% are certified a silver certificate (Comendador, Valdes & Lisker, 2019).

The studies have investigated the effective actions in reducing the indirect GHG2 emissions by airlines are very limited (e.g. Migdadi, 2018, 2019 & 2020). According to Migdadi (2018). The significant actions taken by the airlines to reduce the indirect GHG2 emissions are the vehicles and the engine operating actions. The significant actions to reduce the indirect GHG2 emissions as adopted by the low to moderate emphasizer airlines strategic group were the engine design, the operating of engine, and the accreditation according to LEED standards. However, the significant actions adopted by the low emphasizer airlines strategic group were the energy saving action, the using of sustainable energy, and the upgrading of facilities (Migdadi, 2019).

METHODOLOGY AND METHODS

This part of the chapter discussed how the strategic patterns were identified, and how the effectiveness of these patterns were assessed. In this regards, the design of sample was discussed. Also, how the data was collected and from which sources was discussed. Then, how the data was analyzed to formulate strategy patterns, and track the effectiveness of these patterns was discussed in details.

Sample Design

Table1 shows the distribution of sample between regions. It can be seen that the operations of facilities and buildings of six airlines from each region of North America and Europe were analyzed. Four airlines from Asia, one airline from the Middle East, and one airline from South America were surveyed. The airlines facilities and buildings operations were surveyed for 18 airlines. The sample of study was a convenient sample according to the availability of secondary published data.

Sources of Data

The data was collected from the annual sustainability report of airlines. The reports were retrieved from Global Reporting Initiatives (GRI) database. The GRI is an independent international organization established in 1997 in the US. It seeks to help organizations to report their impact on sustainability. It

Table 1. Sample of study

Regions	Facilities and Buildings Operations
NA: North America	6
E: Europe	6
A: Asia	4
ME: Middle East	1
SA: South America	1
Total	18

develops standards of reporting sustainability and publishes reports via its website (GRI, 2017).

Definition and Measures of Variables

The dependent variable of this study was the indirect GHG2 emissions and the independent variables were the building and the facilities operations actions. The indirect GHG2 emissions were defined as the metric tons of indirect CO2 emissions of electricity, power and heat from the direct billing of owned or leased facilities. However, the facilities and the building operation actions were classified into the facilities and the building energy saving, the facilities

Table 2. Indirect GHG2 Emissions Facilities and Buildings Operations Actions

Actions Dimensions	Actions
Facilities and building energy saving	Air-conditioning energy saving
	Installing electricity meters points to allow real reporting and auditing
	Light energy saving
Facilities and building accreditation	Certification of buildings according to LEED standards
Facilities and building sustainable energy	Using sustainable energy at ground operations
	Using heat pumps with aquifers to generate sustainable energy
	using solar panels
Upgrading and replacing facilities	Upgrading the copiers, fax and printer to energy-friendly ones
	Upgrading and adjusting building control and automation system to conserve electricity
	Replacing power equipment at several radar sites
	Upgrading lifts to more efficient models

(Source: Migdadi, 2018, 2020)

and the building accreditation, the facilities and the building sustainable energy, and upgrading and replacing of the facilities. Each dimension of the facilities and the building operations' actions was investigated by a set of actions. The list of actions is presented in Table 2.

Data Analysis

How to assess the indirect GHG2 emissions performance

Phase1: Computing the percentage of change in indirect GHG2 emissions. The following steps were taken to achieve this phase:

1. Recording the actual achieved performance in 2013 and 2016.
2. Computing the percentage of change by using the following formula:

[(GHG2 emissions 2016- GHG2 emissions 2013)/ GHG2 emissions 2013] X100%

Table 3. Rating Scale of Indirect GHG2 Emissions Performance

0	1	2	3	4	5
	Extremely low	Low	Moderate	High	Extremely high

Phase 2: Transform the percentage of change in indirect GHG2 emissions into rating value as presented in Table 3 below. The following steps were taken to achieve this phase:

1. Identifying the airline achieved the best performance.
2. Dividing the performance achieved by each airline by the best performance. The result is the relative performance
3. Multiplying the relative performance by 5 (the maximum rating value of the scale). The result is the rating value.

Phase 3: Ranking the cases according to achieved rating value. The following steps were taken to achieve this phase:

1. Classifying the scale into three rating levels as following:

1. Step 1: identifying the turning value as following

[(5-0)]/3= 1.67

Step 2: Identifying the levels by adding the turning value as following: From more than 0 to 1.67 is low level. From more than1.67 to 3.33 is moderate level. From more than 3.33to 5.00 is high level.

2. Assigning the level for each case according to achieved rating value. The Assigned level represents the rank of the case performance.

How to identify the effective indirect GHG2 emissions strategic patterns

Phase 1: Computing the actions indices of indirect GHG2 emissions. The following steps were taken to achieve this phase:

1. Identifying the actions categories and the actions lists. This study adopted the same classification of Migdadi (2018, 2020).
2. Computing the number of actions taken by each case under each category.
3. Dividing the number of taken actions by each case by the total number of actions could be taken. The result is a relative number called action adoption index.

Phase 2: Transforming the taken action index into rating value according to the rating scale that presented in Table 4 below. The following steps were taken to achieve this phase:

1. Identifying the case achieved the best index value.
2. Dividing the action index achieved by each case by the best value. The result is the relative action taken index.
3. Multiplying the relative index value by 5 (the maximum rating value of the scale). The result is the rating value of taken actions.

Table 4. Rating Scale of Indirect GHG2 Actions Indexes

0	1	2	3	4	5
	Extremely low	Low	Moderate	High	Extremely high

Phase 3: Identifying the relative level of action adoption according to achieved rating value. The following steps were taken to achieve this phase:

1. Classifying the scale into three rating levels as following:
1. Step 1: identifying the turning value as following

[(5-0)]/3= 1.67

Step 2: Identifying the levels by adding the turning value as following: From more than 0 to 1.67 is low level. From more than1.67 to 3.33 is moderate level. From more than 3.33to 5.00 is high level.

2. Assigning the level for each case according to the achieved rating value.

Phase 4: Summarizing the actions categories indices for all cases. The cases were ranked according to the achieved ranking value of the indirect GHG2 emissions.
Phase 5: Identifying the actions categories with a moderate and a high level rating value.
Phase 6: Identifying the cases that took shared actions categories with moderate or high rating value.
Phase 7: Assigning the titles for the cases according to the most shared actions between cases. These titles represent the different adopted patterns. If the patterns were adopted by the cases achieved an improved performance, then, the patterns would be recognized as the effective patterns.

How to compute the impact of actions taken by effective patterns on the achieved indirect GHG2 emissions

Phase 1: Identifying the actions taken by the higher effective cases in comparison with the lower effective and non-effective cases (which were adopted the same pattern). These actions were called the significant actions.
Phase 2: Counting the number of significant actions taken under each action category by the higher effective cases.
Phase 3: Computing the relative impact of each action category by dividing the number of significant action under each category by the total number of significant actions.

Phase 4: Computing the differences in GHG2 emissions between the higher effective and the lower effective or non-effective case as following:

Indirect GHG emission of the higher effective case-indirect GHG2 emission of the lower effective (or non-effective) case

Phase 5: Multiplying the relative impact of each action category by the differences in GHG2 emissions between the higher and the lower effective or non-effective case. The result is the reduction in GHG2 emissions achieved by the action category.

Phase 6: Dividing the reduction in GHG2 emissions achieved under the action category by the number of actions of the category. The result is the amount of reduction in GHG2 emissions achieved by the action.

Phase 7: Computing the average reduction of GHG2 emissions achieved by each action. The result is the average impact of the action under each category.

RESEARCH RESULTS

The aim of this part was to present the effective indirect GHG2 emissions strategies. Accordingly, the ranking of airlines according to the achieved indirect GHG2 emissions, then, the strategic patterns adopted were presented as indicative models. The indicative models include the impact of actions adopted by each strategic pattern to reduce the indirect GHG2 emissions. In order to facilitate the reading of this chapter results Table 5 shows the acronyms used in presenting this study results.

Table 5. List of Acronyms Used in Presenting the Chapter Results

Acronym	Interpretation
A	Facilities and building accreditation
ES	Facilities and building energy saving
GHG2	Indirect greenhouse gas emission
RF	Upgrading and replacing facilities
SE	Facilities and building sustainable energy

Indirect GHG2 Emissions of Airlines

Table 6 shows the indirect GHG2 emissions of Airlines. It can be seen that 56% (10 cases) reduced the indirect GHG2 emissions. The percentages of reducing in the indirect GHG2 emissions were between -0.02% to -24%. The level of reduction according to the rating scale levels was low for five cases, or about 50% of the reduced emissions cases. However, three cases, about 30% of the reduced emissions cases achieved a moderate level. Two cases, about 20% of reduced emissions cases achieved a high reduction in the indirect GHG2 emissions. The percentage of increase in the indirect GHG2 emissions was between 1% to 26%. The level of increase according to the rating scale level was low for four cases, about 50% of the increased emissions cases. Two cases achieved a moderate increase in the indirect GHG2 emissions,

Table 6. Indirect GHG2 Emissions of Airlines Facilities and Building Operations

Case ID	*% of change in emissions*	*5-points rating of change in emissions*	*Rating level*
Reduced Indirect GHG2 Emissions Group			
A1	-24%	4.90	+H
NA2	-23%	4.79	+H
NA1	-17%	3.53	+M
E2	-16%	3.35	+M
A4	-10%	2.08	+M
NA6	-4%	0.89	+L
E5	-4%	0.79	+L
A6	-3%	0.68	+L
E1	-2%	0.44	+L
NA3	-2%	0.38	+L
Increased Indirect GHG2 Emissions Group			
A3	+1%	-0.24	-L
E4	+2%	-0.32	-L
E6	+6%	-1.25	-L
NA5	+10%	-1.98	-L
SA1	+11%	-2.31	-M
NA4	+16%	-3.29	-M
ME2	+24%	-4.92	-H
E3	+26%	-5.00	-H

L: Low **M:** Moderate **H:** High
+: Improvement in the performance (Reduction in the indirect GHG2 emissions)
-: Deterioration in the performance (Increase in the indirect GHG2 emissions)

about 20% of the increased emissions cases and two cases achieved a high increase in the indirect GHG2 emissions.

Effective Indirect GHG2 Emissions Strategic Patterns

Table 7 Shows the actions indices rating values and the strategy patterns of the facilities and the buildings operations. It can be seen that the levels of realized actions indices were moderate and high. For the purpose of this study, only the moderate and the high-level actions indices were recognized as the significant taken actions by the airlines. It can be seen that some

Table 7. Actions Indices and Strategy Patterns of Facilities and Building Operations

level of % of change in Indirect GHG2 emissions	Case ID	Facilities and building energy saving (ES)	Facilities and building accreditation (A)	Facilities and building sustainable energy (SE)	Upgrading and replacing facilities (RF)	Strategy Pattern	Strategy pattern title
Reduced indirect GHG2 Emissions Group							
+H	A1	2.5 (M)	5 (H)	5 (H)	0	ES-A-SE	Energy Saving+
+H	NA2	2.5 (M)	5 (H)	0	0	ES-A	Energy Saving+
+M	NA1	0	0	0	1.7 (M)	RF	Replacing Facilities
+M	E2	0	0	0	0	-------	-------
+M	A4	5 (H)	0	0	0	ES	Energy Saving+
+L	NA6	0	5 (H)	0	1.7 (M)	A-RF	Accreditation+
+L	E5	0	0	5 (H)	5 (H)	SE-RF	Sustainable Energy+
+L	A6	0	0	0	0	-------	-------
+L	E1	5 (H)	0	5 (H)	1.7 (M)	ES-SE-RF	Energy Saving+
+L	NA3	0	5 (H)	0	0	A	Accreditation+
Increased Indirect GHG2 Emissions Group							
-L	A3	0	0	0	0	-------	-------
-L	E4	0	0	0	0	-------	-------
-L	E6	0	0	0	0	-------	-------
-L	NA5	0	0	5 (H)	0	SE	Sustainable Energy+
-M	SA1	0	0	0	0	-------	-------
-M	NA4	0	0	0	0	-------	-------
-H	ME2	0	0	0	0	-------	-------
-H	E3	0	0	0	1.7 (M)	RF	Replacing Facilities

(L): Low **(M)**: Moderate **(H)**: High

cases took a combination of different significant actions, as A1, NA2 and NA6. Other cases took only one action as NA1, A4 and NA3. On the other hand, the significant actions were taken by the reduced emissions and the increased emissions cases. The significant actions taken by the surveyed cases were used as a source to identify the strategic patterns. Accordingly, the last column of Table 7 shows these patterns. Some patterns are more closer than other in term of the shared significant actions. Each pattern has a title according to the most shared actions between the cases. The patterns adopted by the airlines were the energy saving+ pattern, which was adopted by three cases, all of the cases were belong to the reduced emissions group. Facilities replacing pattern, which was adopted by one case belong to the reduce emissions group, and another case belong to the increased emissions group. Accreditation+ pattern, which was adopted by two cases belong to the reduced emissions group. Sustainable energy+ pattern, which was adopted by one case belong to the reduced emissions group, and another case belong to the increased emissions group.

Effective Energy Saving+ Strategic Patterns

Table 8 shows the effective design+ strategic patterns. Four patterns were adopted. The first pattern was adopted by the case A1. This pattern includes the actions categories of the energy saving, the accreditation, and the sustainable energy. The second pattern was adopted by the case NA2. The second pattern includes the actions categories of the energy saving and the accreditation. The third pattern was adopted by A4. This pattern includes the energy saving action category. However, the fourth pattern which was adopted by E1 includes the energy saving, the sustainable energy and the replacing of facilities. The most effective pattern in term of the total impact was the second pattern (ES-A), which was adopted by NA2, the total reduction of the indirect GHG2 emissions was -20.25%. Followed by the first pattern (ES-A-SE), which was adopted by A1. The total reduction of emissions was -15.5%. The third effective pattern was (ES-SE-RF). This pattern was adopted by E1. The least effective pattern was (ES); this pattern was adopted by A4. The total reduction of emissions was -8%.

The impact of each action category on reducing the indirect GHG2 emissions varies for the first pattern (which was adopted by A1). The most effective action category was the accreditation of building and facilities by an impact of -6.97%. The second effective action category was the using of

Table 8. Effective Energy Saving+ Strategic Patterns

Case ID	Strategy Pattern	Actions category Total Impact	Significant actions taken	Average Relative % of change in Indirect GHG2 emissions
A1	ES-A-SE	**Energy Saving (ES)** Total Impact= -3.72%	Installing electricity meters points to allow real reporting and auditing	-3.72%
		Accreditation (A) Total Impact= -6.97%	Certification of buildings according to LEED standards	-6.97%
		Sustainable Energy (SE) Total Impact= -4.81%	Using solar panels	-4.81%
Total Impact of the Pattern				-15.5%
NA2	ES-A	**Energy Saving (ES)** Total Impact= -6.50%	Light energy saving	-6.50%
		Accreditation (A) Total Impact= -13.75%	Certification of buildings according to LEED standards	-13.75%
Total Impact of the Pattern				-20.25%
A4	ES	**Energy Saving (ES)** Total Impact= -8.0%	Installing electricity meters points to allow real reporting and auditing	-8.0%
Total Impact of the Pattern				-8.0%
E1	ES-SE-RF	**Energy Saving (ES)** Total Impact= -6.0%	Air-conditioning energy saving	-3.0%
			Light energy saving	-3.0%
		Sustainable Energy (SE) Total Impact= -3.0%	Using heat pumps with aquifers to generate sustainable energy	-3.0%
		Replacing Facilities (RF) Total Impact= -3.0%	Upgrading the copiers, fax and printer to energy-friendly ones	-3.0%
Total Impact of the Pattern				-12.0%

sustainable energy by an impact of -4.81%. The third was the using of energy saving by an impact of -3.72%. The impact of actions category on reducing the direct GHG2 emission for the second strategic pattern (which was adopted by A7) also varies. The most effective action category was the accreditation of buildings and facilities by an impact of -13.75%. Followed by the energy saving by an impact of -6.5%. The two strategy patterns are sharing the same ranking of the shared action categories in term of reducing the indirect GHG2 emissions, but the impact of the second strategic pattern was more than the first pattern. The impact of the energy saving action category on reducing the indirect GHG2 emission by the third strategic pattern (which was adopted by A4) was -8%. The impact of each action category on reducing the indirect GHG2 emissions varies for the fourth strategic pattern (which was adopted by E1). The most effective action category was the energy saving by an impact of -6%. Followed by using the sustainable energy and replacing of facilities by a reduction of indirect GHG2 emission by -3% for each action category.

One action related to each action category was taken by the first pattern (which was adopted by A1). The most effective action taken by the first pattern was the certificating of buildings according to LEED standards by an impact of -6.9%. The next effective action was the using of solar panel by an impact of -4.81%. One action related to each action category was taken by the second strategy pattern (which was adopted by NA2). The most effective action taken by the second pattern was the certifying of buildings according to LEED standards by an impact of -13.75%. The third pattern adopted one action, which was installing the electricity meters points to allow a real reporting and auditing by an impact of -8%. The fourth strategic pattern (which was adopted by E1) took different numbers of actions. Two actions were related to the energy saving category and one action was related to the sustainable energy and the replacing of facilities. The impact of all action was the same by an impact of -3% reduction of the indirect GHG2 emissions for each action.

There was some shared actions between the strategic patterns. The first shared action was installing the electricity meters points to allow a real reporting and auditing. This was a shared action between the first and the third strategic pattern. It was the more effective as adopted by the fourth pattern by an impact of -8%. The second shared action was the accrediting of buildings according to LEED standards. This action was a shared action between the first and the second pattern, but it was more effective as adopted by the second pattern by an impact of -13.75%. The third shared action was between the second and the fourth strategic pattern. This action was the light energy saving. It was more effective as adopted by the second strategy pattern by -6.5% reduction in the indirect GHG2 emissions.

Effective Replacing Facilities Strategic Patterns

Table 9 shows the effective replacing facilities strategic patterns. Only one strategic pattern was adopted by the case NA1. In addition, one action was taken by NA1, this action was upgrading the copiers, the fax and the printer to energy-friendly ones. The impact of this action on reducing the indirect GHG2 emissions was -43%.

Table 9. Effective Replacing Facilities Strategic Patterns

Case ID	Strategy pattern	Action category Total Impact	Significant actions taken	Average Relative % of change in Indirect GHG2 emissions
NA1	RF	**Replacing Facilities (RF)** Total Impact= -43%	upgrading the copiers, fax and printer to energy-friendly ones	-43%

Effective Accreditation⁺ Strategic Patterns

Table 10 shows the effective Accreditation+ strategic patterns. Two strategic patterns were adopted. The first pattern was adopted by the case NA6. This pattern includes the actions categories of accreditation and replacing facilities. The second pattern was adopted by the case NA3. The second pattern includes the action category of the accreditation of building and facilities. One shared actions category between the two strategic patterns which was the accreditation of building and facilities. The impact of each action category in reducing the indirect GHG2 emissions was equal for the first pattern (which was adopted by NA6). The impact of each action category was -1%. The impact of the action category on reducing the indirect GHG2 emissions for the second strategic pattern (which adopted by NA31) was -11.7%. One shared action between the two patterns, this action was certificating of buildings according to LEED standards. Its impact was more effective as adopted by the second pattern by an impact of -11.7%.

Table 10. Effective Accreditation+ Strategic Patterns

Case ID	Strategy pattern	Action category Total Impact	Significant actions taken	Average Relative % of change in Direct GHG2 emissions
NA6	A-RF	**Accreditation (A)** Total Impact= -1.0%	Certification of buildings according to LEED standards	-1.0%
		Replacing Facilities (RF) Total Impact=-1.0	Upgrading and adjusting building control and automation system to conserve electricity	-1.0%
Total impact of the pattern				-2.0%
NA3	A	**Accreditation (A)** Total Impact= -11.7%	Certification of buildings according to LEED standards	-11.7%
Total impact of the pattern				-11.7%

Effective Sustainable Energy⁺ Strategic Patterns

Table 11 shows the effective sustainable energy+ strategic pattern. One strategic pattern was adopted by the case E3. This pattern includes the actions categories of sustainable energy and replacing of facilities. The impact of each action category on reducing the indirect GHG2 emissions varies. The most effective action category was replacing of facilities by an impact of -9.75%. Three actions related to replacing of facilities were taken. These actions were upgrading the copiers, the fax and the printer to energy-friendly ones, upgrading and adjusting the building control and the automation system to conserve electricity, and replacing the power equipment at several radar sites. The impact of these three actions was equal, which was -3.25% reduction in the emissions. Only one action was taken related to the sustainable energy, which was using the solar panel, by an impact of -3.25%.

Summary of Effective Indirect GHG2 Emissions Strategic Patterns

Table 12 shows a summary of the effective indirect GHG2 emissions strategic patterns. It can be seen that the most effective pattern was replacing facilities

Table 11. Effective Sustainable Energy+ Strategic Patterns

Case ID	Strategy pattern	Action category Total Impact	Significant actions taken	Average Relative % of change in Indirect GHG2 emissions
E5	SE-RF	**Sustainable Energy (SE)** Total Impact=-3.25%	Using solar panel	-3.25%
		Replacing Facilities (RF) Total Impact= -9.75	Upgrading the copiers, fax and printer to energy-friendly ones	-3.25%
			Upgrading and adjusting building control and automation system to conserve electricity	-3.25%
			Replacing power equipment at several radar sites	-3.25%
Total impact of the pattern				-13.00%

(RF), which was adopted by NA1. The impact of this pattern on reducing the indirect GHG2 emissions was high by -43%. The next ranked patterns, which had a moderate impact on reducing the indirect GHG2 emissions were energy saving+ pattern (ES-A) and energy saving+ pattern (ES-A-SE). The impact of these two patterns was -20.25% and -15.50% respectively. The last ranked patterns, which had a low impact on reducing the indirect GHG2 emissions were sustainable energy + pattern (SE-RF), energy saving+ pattern (ES-SE-RF), accreditation+ (A), energy saving+ (ES), and accreditation+ (A-RF). The impact of these patterns was -13%, -12%, -11.70%, -8.00%, and -2.00% respectively.

DISCUSSION

The first energy saving+ strategy pattern (which was adopted by A1) shared two actions categories with the second energy saving+ pattern (which was adopted by NA2). These actions categories were the accreditation of buildings and facilities, and the energy saving. The second patterns was the most effective in these two actions categories. This could be related to adopting

Table 12. Summary of Airline Effective Indirect GHG2 Emission Strategic patterns

Case ID	Strategy pattern title	Strategy Pattern	Relative average % of change in Indirect GHG2 emissions	Facilities and building energy saving (ES)	Facilities and building accreditation (A)	Facilities and building sustainable energy (SE)	Upgrading and replacing facilities (RF)
NA1	RF	Replacing Facilities	-43% (H)	------	------	------	-43%
NA2	ES-A	Energy Saving+	-20.25% (M)	-6.50%	-13.75%	------	-----
A1	ES-A-SE	Energy Saving+	-15.50% (M)	-3.72%	-6.97	-4.81%	-----
E5	SE-RF	Sustainable Energy+	-13.00% (L)	------	------	-3.25%	-9.75%
E1	ES-SE-RF	Energy Saving+	-12.00% (L)	-6.00%	-----	-3.00%	-3.00%
NA3	A	Accreditation+	-11.70% (L)	------	-11.70%	------	------
A4	ES	Energy Saving+	-8.00% (L)	-8.00%	-----	-----	-----
NA6	A-RF	Accreditation+	-2.00% (L)	-1.00%	------	------	-1.00%

(L): Low **(M)**: Moderate **(H)**: High

different actions categories of LEED standards. The standards are classified into four categories: certified, silver, gold and platinum (Comendador, Valdes & Lisker, 2019). The second case pattern (A7) could adopt a higher ranked category than the pattern of A1. In addition, LEED standards does not recommend a particular environmental conservation producers or actions. Accordingly, each airline can adopt different environmental actions. On the other hand, the energy saving actions that adopted by the second strategy pattern differ from the first pattern. The second pattern adopted the action category of using the energy saving lighting. In the airports, the second source that consumes about 19.82% of the energy was light (Ortega Alba & Manana, 2016), accordingly, the airports directed toward using LED lights (Budd, Budd & Ison, 2015; Malik, 2017; Jun & Jie, 2018). The using of LED lights in airports was assumed to save energy by about 70%-90% (Baxter, Srisaeng & Wild, 2018b).

The third energy saving+ strategy pattern (which was adopted by A4) took only one energy saving action. This action was installing the electricity meters points to allow a real reporting and auditing of the consumed electricity. This action was shared with the first strategy pattern, but the action was more effective as adopted by the third pattern in comparison with the first pattern. This could be related to the number of buildings and facilities have the meters points. The more buildings have the electricity meter points, the more accurate auditing and reporting of the consumed electricity. Moreover, this depends on the kind of electricity meter used for this purpose. The types of electricity meters are flat rate, interval meter, smart meter, and solar meter. Each meter differs in term of how to count the usage of electricity and how to manage the electricity usage. The most advanced one, which is the smart meter, allows knowing when to adjust the usage in order to save the energy and costs (Electricity and Gas Comparison, 2019).

Two sustainable energy actions were taken. One action was taken by the first energy saving + strategy pattern (which was adopted by A4), this action was using the solar panel. The another action was using the heat pumps with aquifers to generate the sustainable energy, this action was taken by the fourth energy saving+ strategic pattern (which was adopted by E1). The most effective action was using the solar panel. Solar panel is the most recommended sustainable source of the energy for the airports by many researchers (e.g. Budd, budd, & Ison, 2015; Sukumaran & Suhakar, 2017; Baxter, Srisaeng & Wild 2018b; Baxter, Srisaeng & Wild, 2019;). The solar panels could reduce the 2GHG emissions of the airports by 12,569 tons (Xu & Deng, 2018) or 59,200 tons (Sukumaran & Suhakar, 2017).

Two energy saving actions were taken by the fourth energy saving+ strategy pattern (which was adopted by E1). These actions were using air-conditioning energy saving and using energy saving lighting. The most intensive facilities for the energy consumptions at airports are HVAC (Heat, Ventilation, Air Conditioning) system and the lighting systems (Ortega Alba & Manana, 2017). HAVC consumes about 24.5% of the energy. However, lighting consumes about 19.82% of the energy (Ortega Alba & Manana, 2016). However, the effectiveness of these two actions was less than the using of saving lighting of the second energy saving+ pattern (adopted by NA2). The effectiveness of saving lighting depends on whether these lights are replaced everywhere (Budd, Budd & Ison, 2015; Baxter, Srisaeng & Wild, 2018a). Moreover, different actions could be adopted to have more energy saving HAVC. For example, using the automated energy-saving air conditioning, and linking on-off function of AC equipment with the airport passengers IS. This action reduced 1500 tons of the emissions a year (Baxter, Srisaeng & Wild, 2018b). Alternatively, the adopted action could be renovate the heating/cooling modifications and the pipe network (Jun & Jie, 2018).

The only action was taken by replacing facilities strategy pattern (which was adopted by NA1) was the upgrading of copiers, fax and printer to energy-friendly ones. This action was the most effective action adopted by all strategic patterns in reducing the indirect GHG2 emissions, since the electrical equipment such as the printers, the photocopiers, the desktops supplied in almost all departments and units, also widely used for day-to-day operations (Zaini, Mahyudin, Saleh & Ahmad, 2019). So different detailed actions could be adopted as using more upgraded facilities that has timer switches to turn off such equipment automatically outside office hours. This technology could save up to 67% of its daily energy consumption if currently left on 24 hours a day. In addition, using networked multi-function devices (MFDs) such as combined printers/copiers can significantly reduce the energy consumption (Asabere, Acakpovi & Quaynor, 2016).

Two accreditation+ strategic patterns were adopted. The most effective pattern was that adopted by NA3. Only one actions was taken by this pattern, which was the certification of buildings according to LEED standards. This action was shared with the first accreditation+ strategy pattern that was adopted by NA6. This could be related to the higher rank of accreditation adopted by NA3. The higher ranked LEED standard such as silver, gold or platinum required more adoption of the actions related to the sustainable sites, the energy actions, and the indoor environmental quality (US Green Building Council, 2019).

One sustainable energy+ pattern was adopted. All actions adopted by this pattern were equal in their effectiveness. However, replacing facilities actions category was more effective than using the sustainable energy. The copiers, the fax and the printer were upgraded to the energy-friendly ones. Also, upgrading and adjusting building control and automation system to conserve electricity and replacing power equipment at several radar sites. As discussed earlier the printers, the photocopiers, the desktops supplied in almost all departments and units, also widely used for day-to-day operations (Zaini, Mahyudin, Saleh & Ahmad, 2019), so it should be upgraded to be more energy efficient equipment.

The automation system to conserve electricity, which is also known, smart electricity grid, could be realized by many applications as using the smart meters. The smart meters record and report the electricity consumption information automatically to the utility service provider over the wires or wirelessly, and either also, it can be remotely connected or disconnected. In addition, adopting the smart appliances could reduce the energy usage, these devices are communicating directly with the grid to indicate how much energy they are consuming, accordingly, and it is automatically adjusting its consumption of the energy (Gavoukian, Polonetsky & Wolf, 2010).

CONCLUSION

The aim of this chapter was to report the effective airlines green operations strategy in reducing the indirect GHG2 emissions. This research issue has been raised by limited studies. Most of the previous studies have reported the energy saving actions of the airports. The main results of this study were; about 56% (10 cases) reduced the indirect GHG2 emissions. The percentages of reducing in the indirect GHG2 emissions was between -0.02% to -24%. The level of reduction according to the rating scale levels was low for five cases, about 50% of reduced emissions cases. Eight effective strategy patterns were adopted. Four effective energy saving+ strategy patterns were adopted. The most effective pattern was (ES-A). One effective replacing facilities strategy pattern was adopted (RF). This pattern was the most effective pattern in reducing the indirect GHG2 emissions. Two accreditation+ strategy patterns were adopted. The most effective pattern in reducing the indirect GHG2 emissions was (A), this pattern adopted only one action, which was the accreditation of buildings and facilities according to LEED standards. One sustainable energy+ strategy pattern was adopted (SE-RF), the most effective action category of this pattern was the replacing of facilities.

The results of this chapter will benefit the academics and the industry experts alike. The academic have in-depth reporting for the relationship between the actions and the indirect GHG2 emissions. Therefore, the academics can develop hypotheses for the future researches. The experts in the industry have a clear idea about the different strategy patterns could be adopted. In addition, the required actions and their impact in reducing the indirect GHG2 emissions. Despite the significant contribution of this chapter, some recommendations could be given for the future researchers. It is recommended to investigate a larger sample size to have more valid results. It is recommended to investigate the interrelationship between the actions taken to have a better understanding for the independent, the moderating, the mediating and the dependent variables.

REFERENCES

Asabere, N. Y., Acakpovi, A., & Quaynor, N. (2016). Encouraging Green ICT implementation strategies in polytechnic education in Ghana. *World, 6*, 7.

Baxter, G., Srisaeng, P., & Wild, G. (2018a). Sustainable airport energy management: the case of Kansai international airport. *IJTTE. International Journal for Traffic and Transport Engineering*, *8*(3), 334–358. doi:10.7708/ijtte.2018.8(3).07

Baxter, G., Srisaeng, P., & Wild, G. (2018b). An assessment of airport sustainability, Part 2—Energy management at Copenhagen Airport. *Resources*, *7*(2), 1–27. doi:10.3390/resources7020032

Baxter, G., Srisaeng, P., & Wild, G. (2019). Environmentally sustainable airport energy management using solar power technology: The case of Adelaide Airport, Australia. *IJTTE. International Journal for Traffic and Transport Engineering*, *9*(1), 81–100. doi:10.7708/ijtte.2019.9(1).07

Budd, T., Budd, L., & Ison, S. (2015). Environmentally sustainable practices at UK airports. *Proceeding of the ICE-Transport*, *168*(2), 116-123.

Cavoukian, A., Polonetsky, J., & Wolf, C. (2010). Smart privacy for the smart grid: Embedding privacy into the design of electricity conservation. *Identity in the Information Society*, *3*(2), 275–294. doi:10.100712394-010-0046-y

Comendador, V. F. G., Valdés, R. M. A., & Lisker, B. (2019). A Holistic Approach to the Environmental Certification of Green Airports. *Sustainability*, *11*(15), 1–38.

Electricity and Gas Comparison. (2019). *Electricity Meter Types*. https://electricityandgas.com.au/guide/electricity-meter/

Ferrulli, P. (2015). Green airport design evaluation—Method and tools. Looking to methods and tools for the research in design and architectural technology. *Transportation Research Procedia*, *14*, 3781–3790. doi:10.1016/j.trpro.2016.05.463

GRI. (2017). *About GRI*. https:// www.globalreporting.org/Information/about-gri/Pages/default.aspx

Jun, T. L., & Jie, C. J. (2018). Research on China Airport Performance Evaluation of Energy Conservation and Emission Reduction Based on Data Envelopment Analysis Model. In *IOP Conference Series: Earth and Environmental Science* (Vol. 189, No. 5, p. 052011). IOP Publishing. 10.1088/1755-1315/189/5/052011

Malik, K. (2017). Assessment of Energy Consumption Pattern and Energy Conservation Potential at Indian Airports. *Journal of Construction in Developing Countries*, *22*(suppl. 1), 97–119. doi:10.21315/jcdc2017.22.supp1.6

Migdadi, Y. K. A. A. (2018). Identifying the best practices of airlines' green operations strategy: A cross-regional worldwide survey. *Environmental Quality Management*, *28*(1), 21–32. doi:10.1002/tqem.21575

Migdadi, Y. K. A. A. (2019). Identifying the effective taxonomies of airline green operations strategy. *Management of Environmental Quality*.

Migdadi, Y. K. A. A. (2020). Green Operational Strategy for Airlines: Content and Regional Analysis. In Cases on Green Energy and Sustainable Development (pp. 193-229). IGI Global.

Ortega Alba, S., & Manana, M. (2016). Energy research in airports: A review. *Energies*, *9*(5), 1–19. doi:10.3390/en9050349

Ortega Alba, S., & Manana, M. (2017). Characterization and analysis of energy demand patterns in airports. *Energies*, *10*(1), 1–35. doi:10.3390/en10010119

Sotos, M. (2015). *GHG Protocol Scope 2 Guidance. An amendment to the GHG protocol corporate Standard.* World Resources Institute (WRI), in association with the World Business Council for Sustainable Development (WBCSD).

Sukumaran, S., & Sudhakar, K. (2017). Fully solar powered Raja Bhoj International Airport: A feasibility study. *Resource-Efficient Technologies*, *3*(3), 309–316. doi:10.1016/j.reffit.2017.02.001

UN-Secretary-General-High-Level Advisory Group on Sustainable Transportation. (2015). *Mobilizing sustainable transport for development*. https://sustainabledevelopment.un.org/content/documents/2375Mobilizing%20Sustainable%20Transport.pdf

US Green Building Council. (2019). *About LEED*. https://new.usgbc.org/press/about-leed

Xu, J. K., & Deng, Y. W. (2018). Research on Application of Renewable and Clean Energy in Beijing New Airport. *DEStech Transactions on Environment, Energy and Earth Sciences*.

Zaini, Z. H., Mahyudin, N. K. B., Saleh, M. K. M., & Ahmad, A. (2019). Electricity consumption: A case study of energy efficiency program among kuim. *Journal of Engineering and Health Sciences*, *2*(1), 1–8.

Chapter 4
Airline Effective Waste Management Strategies

ABSTRACT

This chapter reports the effective wastes management strategies. To achieve this aim, the waste management practices of 14 airlines from four regions were investigated. The data used in this chapter was a secondary published data in the annual sustainability reports. The data was retrieved from GRI (Global Reporting Initiatives) website. The special tailored data analysis techniques were used for this purpose. The chapter reported four on-board and ground strategy patterns and three hybrid patterns. Hybrid strategy patterns were better ranked than the most of on-board and ground strategy patterns. This chapter helps the decision makers and the academics alike, since the impact of each action under each action category of each pattern on wastes reduction was reported.

INTRODUCTION

The management of wastes generated from the flight operations is one of the green airlines management issues. The cabin wastes generated per year is about 5.2 million tons. It costs about $ 500 million, and it is expected to be doubled until 2030 (Reiss & Becker, 2018). Each passenger generates on average 1.4 k.g. of wastes per trip. Despite the early concern about managing the cabin wastes, until now the most airlines and the catering companies have been recycled very little, and the wastes obtained is typically of a low

DOI: 10.4018/978-1-7998-4255-2.ch004

quantity due to the mix of multiple wastes fractions. There is some changes in the trend of wastes recycling over the last few years, but the efforts are still single initiatives, lacking comprehensive and holistic concern (Blanca-Alcubilla et al., 2018).

In this regard, the airlines should take more effective actions to reduce wastes, there is some actions were taken by the airlines, but the focus should be on the most effective actions. Accordingly, the airlines should have a better insight about the most effective actions to reduce wastes. The waste management in the air transport industry is a field with a wide range of research and study, but has not fully matured (Mehta, 2015). The most of previous studies have focused on reporting the on-board (in-flight) wastes reduction actions (e.g. Blanca-Alcubilla et al., 2018; Tofalli, Loizia & Zorpas, 2018). The actions have been traced by the most studies were limited in their scope. Most of the previous studies have reported the actions with a limited concern about the different strategic patterns adopted. The practices of airlines were investigated for airlines in one country or one region for some selected cases (e.g. Blanca-Alcubilla et al., 2018; Tofalli, Loizia & Zorpas, 2018).

This chapter bridged the gap of the previous studies by reporting the effective green operations strategies in reducing the wastes. This study reported the practices of many airlines in different regions and developed strategic patterns. This chapter adopted the benchmarking methodology by making a comparison and reporting of the most effective patterns. Accordingly, the main question of this chapter was: what were the most effective green strategic patterns that adopted by airlines to reduce the wastes?, to answer the question the following objectives were realized:

1. Ranking airlines according to its wastes reduction performance.
2. Identifying the effective wastes strategic patterns that adopted by airlines.
3. Identifying the significant green actions taken by the effective strategic patterns.
4. Identifying the impact of significant green actions on reducing the wastes.

This chapter is arranged in sections: the first one is the literature review, followed by the methodology and methods, then, the results of the empirical study, next, the discussion, and the final section is the conclusions.

LITERATURE REVIEW

This part of the chapter discussed the key literature of airlines waste management. The first part of the literature review aimed to discuss how the waste indicator is measured, so different methodologies and models were presented, also the factors and assumptions that should be considered in measuring the waste indicator were discussed. The second part of the literature review discussed the green operations actions lists, categories, and the strategy patterns.

Airline Waste Definition and Composite

The airlines wastes could be classified into the on-board (cabin) and the ground wastes, the industrial wastes and the hazard wastes (Migdadi, 2018; 2019; 2020). The composite of cabin wastes includes the clean paper (mainly newspaper) which is the largest component ranging from 32 to 71% of the total wastes weight. Plastic as the drinking cups and the food covers is the next major component which forms 13% to 17% of the total wastes weight. Glasses forms about 4%. Aluminum cans forms about 2% to 4% of the total weight (Li et al., 2003; Hondett, 2010). Accordingly, the most significant waste is paper, in which newspapers, magazines, and advertisements count almost 9.21% to 18.91%, disk paper count 0.96% to 2.65% of the paper waste, the containers of coffee and tea bags count 4.6% to 14.78%, and toilet paper which counts 3.85% to 11.90% of the wastes (Mehta, 2015).

The cabin wastes generated per passenger on the long and medium-haul flights was higher than the short-haul flights. This is because of more meals and other services provided for the longer flights (Li et al., 2003). No differences in the wastes composite according to the classes of passengers (Li et al, 2003). Different origins of flights generate a different composite and quantity of wastes (Tofalli, Loizia & Zorpas, 2018). The concern about managing the wastes is affected by the availability of appropriated facilities, which are rarely available, and the regulations, which are too restrictive (Blanca-Alcubilla et al., 2018)

Industrial wastes are mainly the aircrafts parts generated from the retired, the stored, the abandoned aircrafts, and the aircraft that previously involved in accidents. Over the past 35 years more than 1500 commercial aircrafts have been retired worldwide. On average, about 700 aircraft are being retired each year (IATA, 2018). About 82% of the retired aircrafts are small passenger and about 10% of the aircrafts were retired before 17 years. It is estimated that, 12000 aircrafts will be retired in the next two decades. Industrial wastes management is affected by the regulations and the standards at the international, the regional and the local level as the regulations of aircraft retirement and ICAO standards (IATA, 2018).

Hazard wastes are substances that because of its quantity, physical, concentration or infectious characteristics, or when its improperly treated, stored, transported or disposed cause a series illness, or pose a substantial or potential hazard to the human health or for the environment. In the aircraft industry, it can originate from the service of repair of aircraft components. These could be classified according to its substance to solid, gas, and liquid. Gas is the most hazard, followed by the liquid, and the least harmful is the solid. Also the hazard wastes could be classified into aircraft systems wastes (as gasoline, fuel, brake, ferrous, mitogen…etc.), aircraft servicing (as dry, spry lubricants, engine cleaners, pints…etc.) and component shops (as argon gas, compressed air…etc.) (Wieher, 2014).

Airlines Waste Management Operational Actions and Strategy Patterns

Airlines requesting a new approach to handle the wastes at the operational and the strategic level (Reiss & Becker, 2018). Airports play a vital role in the aviation wastes management. The efficient and effective wastes management system involves normally the waste minimization (waste avoidance or prevention, waste reuse, recycling, energy recovery) as well as the waste disposal (London Stansted Airport, 2010; Reiss & Becker, 2018; Baxter, Srisaeng & Wild, 2018).

The recycling and the reusing of cabin wastes involves the collection, the sorting and the transferring. The collection could be mixed or not mixed. The collection of wastes could be happened after drinking or eating directly, also prior landing. This could be happened by the cabin crew, by the cleaners, or by the passengers. It is recommended to do the collection by using a green tinted bag, which is slight enough to allow the content of bags to be clearly seen. The sorting of wastes includes loading off the bags of mixed or separated recyclables. This could happen on the ground or during the turnaround cleaning. The cleaners or a catering company could make this sorting. The collected wastes will be transferred to the recycling facilities by a cleaning company toward the correct containers at the waste disposal sites in the airport (Hondett, 2010). Nearly 60% of the wastes generated during flight can be easily recycled (Mehta, 2015).

The management of hazard wastes includes identifying the hazard wastes, the storage of hazard wastes, the disposal of hazard wastes and minimize the hazard waste. The strategies of wastes minimization strategies are recycling, use reuse and prevention (Wieher, 2014). The management of industrial wastes includes decisions as the location of facilities will be used for disassembly, the capacity and the accreditation of facilities, and the marketability of parts-out components in a market. The disassembled aircraft parts will be evaluated, then it will be either recycled, disposed, landfilled (for non-recyclable parts), or the parts could be used for other uses whether by making changes in the parts or without. The valuable parts of the aircrafts could be used again with or without repair. 85% to 90% of the content of retired aircrafts are reused or recycled (IATA, 2018).

The most comprehensive classification of wastes reduction actions was the classification developed by Migdadi (2018, 2020). According to this classification the actions are on-board and ground wastes recycling, upcycling, reusing and reduce using of paper. In addition, the industrial wastes recycling, upcycling and reusing. Moreover, the hazard wastes recycling, upcycling and using. The waste management in the air transport industry is a field with a wide range of research and study, but the majority of these studies have investigated the practices in one country context, or in a limited number of regions. Very limited studies have investigated the effective waste management strategies (e.g. Migdadi, 2018, 2019).

According to Migdadi (2018), one of the effective strategic pattern was the waste management (W.M.). The significant action taken in this regard was the paper saving action. This strategy pattern was adopted by one airline. The study of Migdadi (2019) which is a more in-depth, according to this study, the strategy pattern of moderate to low emphasizer adopted some significant action related to the waste reduction. The significant actions taken were related to on-board and ground wastes recycling, upcycling and reusing. This strategy pattern was adopted by three regions (Asia, Europe and the North America). Despite the more in-depth and clear reporting for the strategy patterns by Migdadi (2018, 2019) studies, these studies have not investigated in-depth the impact of action items, and the reporting was for the general strategy patterns of airlines. Therefore, this chapter bridged the gap by reporting in-depth the effective strategy pattern adopted.

METHODOLOGY AND METHODS

This part of the chapter discussed how the strategic patterns were identified, and how the effectiveness of these patterns were assessed. In this regards, the design of sample was discussed, also, how the data was collected was discussed. Then, how the data was analyzed to formulate the strategy patterns, and track the effectiveness of these patterns was discussed in details.

Sample Design

The sample of study was a convenient sample according to the availability of secondary published data. The airlines waste management practices were surveyed for 14 airlines. Table1 shows the distribution of sample between regions. It can be seen that the waste management practices of five airlines of North America were surveyed, four airlines from each region of Asia and Europe were surveyed. One airline from the Middle East was surveyed.

Table 1. Sample of study

Regions	Waste Management
NA: North America	5
A: Asia	4
E: Europe	4
ME: Middle East	1
Total	14

Sources of Data

The data was collected from the annual sustainability report of airlines. The reports were retrieved from Global Reporting Initiatives (GRI) database. The GRI is an independent international organization established in 1997 in the US. It seeks to help organizations to report their impact on sustainability. It develops standards of reporting sustainability, and publishes reports via its website (GRI, 2017).

Definition and Measures of Variables

The dependent variable was the waste generated, and the independent variables were the waste management actions. Wastes generated were defined as tons of the on-board, the ground, the hazard and the industrial wastes. However, the wastes management actions were classified into different dimensions as recycling, upcycling and reusing of the on-board, the ground, the hazard and the industrial wastes and reducing the used paper. The list of actions is presented in Table 2.

Data Analysis

How to assess the wastes management performance

Phase1: Computing the percentage of change in wastes generated. The following steps were taken to achieve this phase:

1. Recording the actual achieved performance in 2013 and 2016.
2. Computing the percentage of change by using the following formula:

[(Wastes generated 2016- wastes generated 2013)/ wastes generated 2013] X100%

Phase 2: Transforming the percentage of change in wastes generated into rating value as presented in Table 3. The following steps were taken to achieve this phase:

1. Identifying the airlines achieved the best performance.
2. Dividing the performance achieved by each case by the best performance. The result is the relative performance.
3. Multiplying the relative performance by 5 (the maximum rating value of the scale). The result is the rating value of performance.

Phase 3: Ranking the cases according to achieved rating value. The following steps were taken to achieve this phase:

1. Classifying the scale into three rating levels as following:

1. Step 1: identifying the turning value as following:

[(5-0)]/3= 1.67

Table 2. Airline Waste Management Actions

Action category	Actions
On-board and ground waste recycling	Adopting a recycling tracking system
	Auditing internal recycling
	Increasing the number of recycling stations
	Recycling electronic boarding passes
	Recycling paper
	Recycling plastic
	Recycling cans
	Using a video system to encourage passengers to sort their waste
	Having a print smart campaign
	Using the cabin intercom to encourage passengers to sort their waste
On-board and ground wastes upcycling	Using outdated vests to manufacture travel amenity bags and kits
	Transforming tarpaulins into pencil cases
	Modifying scrapped service carts
	Shredding uniforms and using them for furniture blankets
	Converting seats into leather products
Reusing on-board and ground waste	Donating blazers and trench coats to charities
Reducing the use of paper	Using electronic airway bills for freight documents
	Using electronic boarding passes
	Providing crews with iPads instead of newspapers
Recycling industrial waste	Recycling aircrafts parts
	Recycling iron
	Recycling of cargo shrink-wrap
Upcycling industrial waste	Burning for energy recovery
Reusing industrial waste	Keeping aircraft parts in warehouses as spare parts
	Reusing wooden cargo pallets in land scrapping
	Upcycling some parts of seats and converting them into leather seats
	Collecting scraps of material and selling them for qualified companies to reuse
	Reusing carpets
	Keeping aircraft parts in warehouses as spare parts
	Reusing the wooden cargo pallets in landscaping
Recycling hazardous waste	Recycling of hazardous waste
	Recycling maintenance waste
	Inventory and tracking systems to track hazardous material early
Upcycling hazardous waste	Burning it to recover energy
Using hazardous waste	Reducing the use of chemicals
	Collecting aviation fuel used during maintenance for later use

(Source: Migdadi, 2018, 2020)

Step 2: Identifying the levels by adding turning value as following

From more than 0 to 1.67 is low level. From more than1.67 to 3.33 is moderate level. From more than 3.33to 5.00 is high level.

Table 3. Rating Scale of Waste Generated Performance

0	1	2	3	4	5
	Extremely low	Low	Moderate	High	Extremely high

2. Assign the level for each case according to achieved rating value. The Assigned level represents the rank of the case performance.

How to identify the effective wastes management strategic patterns

Phase 1: Computing the actions indices of wastes management. The following steps were taken to achieve this phase:

1. Identifying the actions categories and actions lists. This study adopted the same classification of Migdadi (2018, 2020).
2. Computing the number of taken actions by each case under each category.
3. Dividing the number of taken actions by each case by the total number of actions that could be taken. The result is a relative number called action adoption index.

Phase 2: Transforming the action taken index into rating value as presented in Table 4. The following steps were taken to achieve this phase:

1. Identifying the case achieved the best action index value.
2. Dividing the action index achieved by each case by the best index value. The result is the relative action taken index.
3. Multiplying the relative index value by 5 (the maximum rating value of the scale). The result is the rating value of actions taken.

Phase 3: Identifying the relative level of action adoption according to achieved rating value

1. Classifying the scale into three rating levels as following:

Table 4. Rating Scale of Wastes Reduction Actions Indexes

0	1	2	3	4	5
	Extremely low	Low	Moderate	High	Extremely high

1. Step 1: identifying the turning value as following:

[(5-0)]/3= 1.67

Step 2: Identifying the levels by adding the turning value as following: From more than 0 to 1.67 is low level. From more than1.67 to 3.33 is moderate level. From more than 3.33to 5.00 is high level.

2. Assign the level for each case according to the achieved rating value.

Phase 4: Summarizing the actions categories indices for all cases. The cases were ranked according to the ranking value of achieved wastes performance.

Phase 5: Identifying the actions categories with a moderate and a high level rating value.

Phase 6: Identifying the cases that took shared actions categories, with a moderate or a high level rating value.

Phase 7: Assigning the titles for the cases according to most shared actions between cases. These titles represent the different adopted patterns. If the patterns were adopted by the cases achieved an improvement in the performance, then, the patterns would be recognized as the effective patterns.

How to compute the impact of actions taken by effective patterns on the generated wastes

Phase 1: Identifying the taken actions by the higher effective cases in comparison with the lower effective and non-effective cases (which adopted the same pattern). These actions were called the significant actions.

Phase 2: Counting the number of significant actions taken under each action category of the higher effective cases.

Phase 3: Computing the relative impact of each action category by dividing the number of significant action taken under each category by the total number of significant actions.

Phase 4: Computing the differences in the generated wastes between the higher effective and the lower effective or non-effective case as following:

Wastes generated by the higher effective case–Wastes generated by the lower (or non-effective) case

Phase 5: Multiplying the relative impact of each action category by the differences in the wastes generated. The result is the reduction in wastes achieved by the action category.

Phase 6: Dividing the reduction in wastes generated under the action category by the number of actions of the category. The result is the amount of reduction in wastes achieved by the action.

Phase 7: Computing the average reduction in the wastes achieved by each action. The result is the average impact of the action under each category.

RESEARCH RESULTS

This part of this chapter presents the main results of the empirical study. The aim of this part is to present the effective wastes management strategies. This part presented the ranking of airlines according to the achieved reduction in wastes, then, the strategic patterns adopted were presented as indicative models. The indicative models include the impact of actions adopted by each strategic pattern to reduce the wastes. In order to facilitate the reading of the chapter results, Table 5 shows the acronyms of the strategy patterns alternatives.

Wastes Reduction of Airlines

Table 6 shows the wastes reduction of Airlines. It can be seen that 57% (8 cases) reduced wastes. The percentages of reducing wastes was between -0.85% to -40%. The level of reduction according to the rating scale levels was low for five cases, about 50% of the reduced wastes cases. However, one case, about 12.5% of the reduced wastes cases achieved a moderate level, and three cases, about 37.5% of reduced wastes cases achieved a high reduction in the wastes. The percentage of increase in wastes was between +4.87% to

Table 5. List of Acronyms Used in Presenting the Chapter Results

Acronym	Interpretation
HRU	Hazard wastes reusing
HUP	Hazard wastes upcycling
HRC	Hazard wastes recycling
IRC	Industrial wastes recycling
IRU	Industrial wastes reusing
IUP	Industrial wastes upcycling
ORC	On-board and ground wastes recycling
ORP	On-board and ground reduce using of paper
OUP	On-board and ground wastes upcycling
ORU	On-board and ground wastes reusing

Table 6. Wastes Reduction of Airlines

Case ID	% of change in Wastes	5-points rating of change in wastes	Rating level
Reduced Wastes Cases			
NA1	-40.00%	5.0	+H
NA2	-33.25%	4.2	+H
E2	-27.75%	3.5	+H
NA6	-24.66%	3.1	+M
A5	-6.59%	0.8	+L
E1	-4.83%	0.6	+L
NA5	-1.22%	0.2	+L
E4	-0.85%	0.1	+L
Increased Wastes Cases			
NA3	+4.87%	-0.6	-L
E3	+5.00%	-0.6	-L
A6	+5.92%	-0.7	-L
ME1	+11.82%	-1.5	-L
A3	+12.59%	-1.6	-L
A4	+66.43%	-5.0	-H

L: Low **M:** Moderate **H:** High
+: Improvement in the performance (Reduction in the wastes)
-: Deterioration in the performance (Increase in the wastes)

66.43%. The level of increase according to the rating scale level was low for five cases, about 83% of the increased wastes cases. One case achieved a high increase in wastes.

Airlines Wastes Management Strategic Patterns

Table 7 shows the actions indices and the strategy patterns of the wastes management. It can be seen that the levels of realized actions indices were low, moderate and high. For the purpose of this study, only the moderate and the high-level actions indices were recognized as the significant taken actions by the airlines. It can be seen that some cases took a combination of different significant actions, as NA1, NA2 and NA5. Other cases took only one action as A5, E3 and E4. On the other hand, the significant actions were taken by the reduced and the increased wastes cases. The significant actions taken by the surveyed cases were used as a source to identify the strategic

Table 7. Actions Indices Rating Value and Strategy Patterns of Airline Wastes Management

	Case ID	On-board and Ground Wastes (O)				Industrial Wastes (I)			Hazard Wastes (H)			Strategy patterns	Strategy pattern title
		Recycling (ORC)	Upcycling (OUP)	Reusing (ORU)	Reduce using of Paper (ORP)	Recycling (IRC)	Upcycling (IUP)	Reusing (IRU)	Recycling (HRC)	Upcycling (HUP)	Reusing (HRU		
+H	NA1	2.9 (M)	0.0	0.0	5.0 (H)	0.0	0.0	0.0	0.0	0.0	0.0	ORC-ORP	On-board &Ground
+H	NA2	2.1 (M)	0.0	0.0	2.5 (M)	1.7 (L)	5.0 (H)	5.0 (H)	0.0	0.0	0.0	ORC-ORP- IUP-IRU	Hybrid
+H	E2	2.1 (M)	0.0	0.0	0.0	0.0	0.0	0.0	0.0	0.0	2.5 (M)	ORC-HRU	Hybrid
+M	NA6	0.0	0.0	0.0	0.0	0.0	0.0	0.0	0.0	0.0	0.0	-------	-------
+L	A5	2.1 (M)	0.0	0.0	0.0	0.0	0.0	0.0	0.0	0.0	0.0	ORC	On-board &Ground
+L	E1	0.7 (L)	5.0 (H)	0.0	0.0	0.0	0.0	0.0	0.0	0.0	0.0	OUP	On-board &Ground
+L	NA5	2.9 (M)	2.5 (M)	0.0	0.0	5.0 (H)	5.0 (H)	5.0 (H)	5.0 (H)	0.0	5.0 (H)	ORC-OUP-IRC-IUP-IRU-HRU	Hybrid
+L	E4	2.1 (M)	0.0	0.0	0.0	0.0	0.0	0.0	0.0	0.0	0.0	ORC	On-board &Ground
-L	NA3	0.7 (L)	0.0	0.0	0.0	0.0	0.0	0.0	0.0	0.0	0.0	--------	--------
-L	E3	2.9 (M)	0.0	0.0	0.0	0.0	0.0	0.0	0.0	0.0	0.0	ORC	On-board &Ground
-L	A6	0.7 (L)	0.0	0.0	0.0	0.0	0.0	0.0	0.0	0.0	0.0	-------	--------
-L	ME1	5.0 (H)	0.0	0.0	0.0	0.0	0.0	0.0	0.0	0.0	0.0	ORC	On-board &Ground
-L	A3	2.1 (M)	0.0	0.0	0.0	0.0	0.0	0.0	5.0 (H)	0.0	0.0	ORC-HRC	Hybrid
-H	A4	2.1 (M)	5.0 (H)	0.0	0.0	0.0	0.0	0.0	5.0 (H)	0.0	0.0	ORC-OUP-HRC	Hybrid

(L): Low **(M):** Moderate **(H):** HighEffective On-board and Ground Waste Management Strategic Patterns

patterns of the cases. Accordingly, the last column of Table 7 shows these patterns. Each pattern has a title according to the most shared actions between the close patterns cases. The patterns adopted by the airlines were on-board & the ground wastes management, which was adopted by six cases, two of them were adopted by the increased wastes cases. Hybrid pattern, which was adopted by five cases, two of them were adopted by the increased wastes cases.

Table 8 shows the effective on-board and the ground waste management strategic patterns. Four patterns were adopted. The first pattern was adopted by the case NA1. This pattern includes the actions categories of on-board and the ground wastes recycling, and reduce the using of paper. The second pattern was adopted by the case A5. The second pattern includes the actions category of on-board and the ground wastes recycling. The third pattern was adopted by the case E1. This pattern includes the action category of on-board and the ground wastes upcycling. The fourth pattern was adopted by the case E4. This pattern includes the action category of on-board and the ground wastes recycling. The most effective pattern was that adopted by NA1 (ORC-

Table 8. Effective On-board and Ground Wastes Management Strategic Patterns

Case ID	*Strategy Pattern*	*Actions category* *Total Impact*	*Significant actions taken*	*Average Relative % of change in Wastes*
NA1	On-board and ground	**On-board and Ground Wastes Recycling (ORC)** Total Impact= -40.29%	Auditing internal recycling	-17.79%
			Recycling paper	-7.50%
			Recycling plastic	-7.50%
			Recycling cans	-7.50%
		Reduce using of paper (ORP) Total Impact= -28.81%	Using electronic airway bills for freight documents	-14.41%
			Using electronic boarding passes	-14.41%
Total impact of the pattern				-69.10%
A5	On-board and ground	**On-board and Ground Wastes Recycling (ORC)** Total Impact= -12.00%	Recycling paper	-4.00%
			Recycling plastic	-4.00%
			Recycling cans	-4.00%
Total impact of the pattern				-12.00%
E1	On-board and ground	**On-board and Ground Wastes Upcycling (OUP)** Total Impact=-10.40%	Using outdated vests to manufacture travel amenity bags and kits	-5.20%
			Transforming tarpaulins into pencil cases	-5.20%
Total impact of the pattern				-10.40%
E4	On-board and ground	**On-board and Ground Wastes Recycling (ORC)** Total Impact= -6.00%	Recycling paper	-2.00%
			Recycling plastic	-2.00%
			Recycling cans	-2.00%
Total impact of the pattern				-6.00%

ORP) by a total impact of -69.10%. Followed by the pattern was adopted by A5 (ORC) by a total impact of -12%. The next pattern was (OUP) which was adopted by E1 by a total impact of 10.40%. The least effective pattern was (ORC), which was adopted by E4, by a total impact of -6%.

Three shared actions between the three strategic patterns. These actions were the recycling of paper, cans and plastic. These three actions were adopted by the first, the second and the fourth pattern. The impact of these actions varies. These three actions were the most effective by the first pattern (ORC-ORP). The impact was -7.50% by each action. However, the impact was -4% by each action adopted by the second pattern (ORC). The lowest impact was -2% by each action adopted by the fourth pattern (ORC). The most effective action was the auditing of internal recycling which was adopted by first pattern, by an impact of -17.79. The next effective action was the of using electronic airway bills for the freight documents, and using the electronic boarding pass, by an impact of each action by -14.41%. The actions of on-board and upcycling wastes were the next effective actions, by an impact of -5.20% for each action. These actions were using the outdated vests to manufacture travel amenity bags and kits, and transforming the tarpaulins into pencil cases.

Effective Hybrid Wastes Management Strategic Patterns

Table 9 shows the effective hybrid wastes management strategic patterns. Three patterns were adopted. The first pattern was adopted by the case NA2. This pattern includes the actions categories of on-board and the ground wastes recycling, reduce the using of paper and industrial wastes reusing. The second pattern was adopted by the case NA5. The second pattern includes the actions categories of on-board and the ground wastes recycling, on-board and the ground wastes upcycling, the industrial wastes recycling, and the industrial wastes reusing. The third pattern was adopted by the case E2. The third pattern includes the actions categories of on-board and the ground wastes recycling, and the hazard wastes reusing. One shared action category between the three strategic patterns, this category was on-board and the ground wastes recycling. Further two shared actions between the first and the second strategic pattern, these categories were the industrial wastes recycling and the industrial wastes reusing. The first and the second patterns took more actions categories than the third pattern. The most effective strategy pattern according to the total impact was the third pattern, which was adopted by the case E3, the total reduction in wastes was -64.83%. The next effective strategic pattern was

Table 9. Effective Hybrid Wastes Management Strategic Patterns

Case ID	Strategy pattern	Action category Total Impact	Significant actions taken	Average Relative % of change in Wastes
NA2	ORC-ORP-IRC-IRU	**On-board and Ground Wastes Recycling (ORC)** Total Impact= -22.90%	Recycling paper	-7.63%
			Recycling plastic	-7.63%
			Recycling cans	-7.63%
		Reduce using of paper (ORP) Total Impact=-7.63%	Providing Crew with iPad instead of paper	-7.63%
		Industrial Wastes Recycling (IRC) Total Impact=-7.63%	Recycling of cargo shrink-wrap	-7.63%
		Industrial Wastes Reusing (IRU) Total Impact=-7.63%	Reusing wooden cargo pallets in landscaping	-7.63%
Total impact of the pattern				-45.80%
NA5	ORC-OUP-IRC -IRU	**On-board and Ground Wastes Recycling (ORC)** Total Impact= -18.22%	Auditing internal recycling	-4.56%
			Recycling paper	-4.56%
			Recycling plastic	-4.56%
			Recycling cans	-4.56%
		On-board and Ground Wastes Upcycling (OUP) Total Impact=-4.56%	Converting seats into leather products	-4.56%
		Industrial Wastes Recycling (IRC) Total Impact=-9.11%	Recycling iron	-4.56%
			Recycling of cargo shrink-wrap	-4.56%
		Industrial Wastes Reusing (IRU) Total Impact=-4.56%	keeping aircraft parts in warehouses as spare parts	-4.56%
Total impact of the pattern				-36.44%
E2	ORC-HRU	**On-board and Ground Wastes Recycling (ORC)** Total Impact= -48.08%	Increasing the number of recycling stations	-15.67%
			Adopting a recycling tracking system	-16.75%
			Recycling oils from restaurant	-15.67%
		Hazard Wastes Reusing (HRU) Total Impact= -15.67%	burning it to recover energy	-15.67%
Total impact of the pattern				-64.83%

the first pattern, which was adopted by the case NA2, the total reduction in wastes was -45.80%. The lowest effective pattern was the second pattern, which was adopted by NA5, the total reduction in wastes was -36.44%.

The impact of each action category on reducing the wastes varies for the first pattern (which was adopted by the case NA2). The most effective action category was on-board and the ground recycling by an impact of -22.90%. The impact of remaining action categories was equal by an amount of -7.63%. The impact of action categories on reducing the wastes for the second strategic pattern (which was adopted by NA5) also varies. The most effective action category was on-board and the ground wastes recycling by an impact of -18.25%, followed by the industrial wastes recycling category with an impact of -13.67%. the lowest effective action categories were on-board and the ground wastes upcycling and the industrial wastes reusing by an impact of -4.56% for each action category. The impact of actions category on reducing the wastes for the third strategic pattern (which was adopted by E2) also varies. The most effective action category was on-board and the ground wastes recycling by an impact of -48.08%, followed by the hazard wasted reusing category by an impact of -15.67%.

Three actions related to on-board and the ground wastes recycling were shared between the first and the second pattern. These actions were recycling of paper, plastic and cans. The impact of these three actions on reducing wastes was -7.63% for each action adopted by the fist pattern, and -4.56% for each action adopted by the second pattern. On extra action taken by the second pattern, which was auditing the internal recycling, by an impact of -4.56%. On-board and the ground wastes recycling actions that adopted by third pattern differ from those adopted by other patterns and were more effective. These actions were increasing the number of recycling stations, adopting the recycling and tracking system and recycling oils from restaurants. The impact of these actions on reducing the wastes was -15.67%, -16.75% and -15.67% respectively.

One shared action related to the industrial wastes recycling between the first and second pattern. This action was the recycling of cargo shrink-wrap. This action was also more effective as adopted by the first pattern, the impact was -7.63%. The second pattern adopted recycling of iron, the impact of these action was -36.44% for each action. Two different reusing of the industrial wastes actions were adopted by the first and the second pattern. The first pattern adopted the reusing of cargo wooden pallets in land scrapping by an impact of -7.63%. The action adopted by the second pattern was keeping the aircrafts parts in the warehouses as spare parts, by an impact of -4.56%.

Reduce the using of paper action category was adopted by the fist pattern. The only action adopted was providing the crew with iPad instead of paper, the impact of this action was -7.63%. The only industrial wastes upcycling action taken was converting the seats into leather products. This action was adopted by the second pattern, and its impact was -4.56%. The action related to the reusing of hazard wastes was adopted by the third pattern, this action was burning it to recover the energy, the impact of this action was -15.67%.

Summary of Effective Waste Management Strategic Patterns

Table 10 shows a summary of the effective waste management strategic patterns. It can be seen that two strategic patterns achieved a high relative reduction in wastes. These patterns were on-board and the ground wastes management (ORC-ORP) and the hybrid pattern (ORC-HRU). The relative percentage of change in wastes was -69.10% and -64.83% respectively. The next ranked pattern were those adopted by NA2 and NA5, both of them are hybrid patterns (ORC-ORP-IRC-IRU) and (ORC-OUP-IRC-IRU). The relative impact of these two pattern was moderate. The achieved relative reduction in wastes was -22.90% and -18.22% respectively. The last ranked patterns were three on-board and ground wastes management patterns. These patterns were adopted by the case A5, E1 and E4. The relative impact of these patterns

Table 10. Summary of Airline Effective Waste Management Strategic pattern

Case ID	Strategy pattern title	Strategy Pattern	Relative average % of change in wastes	On-board and Ground Wastes Recycling (ORC)	Reduce using of paper (ORP)	On-board and Ground Wastes Upcycling (OUP)	Industrial Wastes Recycling (IRC)	Industrial Wastes Reusing (IRU)	Industrial Wastes Upcycling (IUP)	Hazard Wastes Reusing (HRU)
NA1	On-board and ground	ORC-ORP	-69.10% (H)	-40.29%	-28.81%	------	------	------	------	------
E2	Hybrid	ORC-HRU	-64.83% (H)	-48.08%	------	------	------	------	------	-15.67%
NA2	Hybrid	ORC-ORP-IRC-IRU	-45.80% (M)	-22.90%	-7.63%	------	-7.63%	-7.63%	------	------
NA5	Hybrid	ORC-OUP-IRC-IRU	-36.44% (M)	-18.22%	------	-4.56%	-9.11%	-4.56%		
A5	On-board and ground	ORC	-12.00% (L)	-12.00%	------	------	------	------	------	------
E1	On-board and ground	ORC	-10.40% (L)	-10.40%	------	------	------	------	------	------
E4	On-board and ground	ORC	-6.00% (L)	-6.00%	------	------	------	------	------	------

(L): **Low** (M): **Moderate** (H): **High**

was low, and the achieved reduction in wastes was -12%, -10.40% and -6% respectively.

DISCUSSION

The shared action category between on-board and the ground wastes strategy patterns was on-board and the ground wastes recycling, this indicates for the importance of on-board wastes recycling. According to Airbus (2018) survey, 27% of the airlines show the importance of cabin waste recycling as a very high. However, 72% of the airlines expecting the importance will be more over the next five years. The shared actions between the three strategic patterns of on-board and the ground strategy patterns were the recycling of paper, cans and plastic. This result is justified by the composite of cabin wastes studies. According to these studies, the clean paper (mainly newspaper) was the largest component, ranging from 32 to 71% of total wastes weight. Plastic as the drinking cups and the food covers is the next major component which forms 13% to 17% of total wastes weight. Glasses about 4%. Aluminum cans forms about 2% to 4% of the total weight (Li et al., 2003; Hondett, 2010), so these items are the most proposed items for recycling.

The impact of recycling of paper, plastic and cans was the most effective adopted actions by the first on-board and the ground strategy pattern (ORC-OPR). This is because this strategy is the only pattern adopts the auditing of internal recycling. The audit will specify the amounts of waste and identify the new opportunities for recycling, reuse, and the waste reduction. Also, this helps in evaluating the effectiveness of waste management over time. Carrying out a waste audit is an important first step in developing or refining a waste management plan. In particular, such information is a crucial to develop a recycling program (International Air Transport Organization, no date). Three hybrid waste management strategic patterns were adopted. All patterns adopted one shared action category which was on-board and the ground wastes recycling. The first and the second strategic patterns adopted also the industrial wastes recycling and reusing. The third pattern adopted the hazard wastes reusing. However, the most effective action category adopted by all strategy pattern was on-board and the ground wastes recycling. This indicates the cabin wastes are more manageable than the hazard and the industrial wastes by the airlines. The hazard and the industrial wastes are mainly managed by the airports, furthermore, the management of these wastes at the airports is generally reliant on the national/local regulation (International Air Transport Organization, no date).

The recycling of wooden cargo pallets was the taken industrial waste reusing action by the first pattern. The reusing of cargo wooden pallets is a very important, not only for the airlines but also for all sectors. For example, nearly 2 billion wooden pallets are currently in the circulation in the US alone, with a majority of them are replaced each year. This consumes an estimated 50 percent of the country's annual hardwood harvest (Larox, 2012). The keeping of used aircraft parts as spare parts was taken by the second strategy pattern. The adoption of this action will keep the disposal cost at minimum and reduce the cost of parts itself. The only hazard waste management adopted was burning it to recover the energy, this will reduce the energy costs and improve the carbon-offset practices. The effective action category adopted by all strategy patterns was on-board and the ground wastes recycling. This result is consistent with the results of Migdadi (2019) study. According to this study, the effective actions are related mainly to on-board wastes recycling, upcycling and reusing. In addition, the result of this study is partially supported by the results of Migdadi (2018) study. According to this study, the significant actions taken was the paper reduction action.

The hybrid strategy patterns better ranked than the on-board and the ground wastes management patterns. This indicates managing more sources of wastes, as on-board, the industrial and the hazard wastes will improve the overall wastes reductions, also, this indicates the wastes generated by the airlines are diverse and require more concern by setting an overall strategy to manage all sources of wastes. Despite on-board and the ground wastes are the most important in term of its quantity, other sources of wastes could be more harmful for human and other creatures life as the hazard wastes. In addition, other wastes as the industrial wastes cause harm and require many years to be disposed in the land.

CONCLUSION

The aim of this chapter was to report the effective airlines green operations strategy in reducing the wastes. Despite the widespread studies of wastes management in the airlines industry, limited studies have reported in details the effective green strategies that reduced the wastes. The main results of this study were about 57% of the airlines have reduced the wastes. In addition, the majority of reduced wastes airlines achieved a low level of reduction.

Different strategy patterns were adopted. On-board and the ground strategy patterns were four. All of these patterns adopted on-board and the ground wastes management action category. The most effective on-board and the ground wastes management pattern was the pattern that adopted actions categories related to on-board and the ground wastes management category, and on-board paper recycling. Three effective hybrid strategy patterns were adopted. All of them adopted on-board and the ground wastes recycling, two of them adopted shared actions categories as the industrial wastes recycling and the industrial wastes reusing. One of them adopted on-board paper recycling. In addition, one of them adopted on-board and the ground upcycling actions, and one of them adopted the hazard wastes reusing. The most effective pattern was that adopted on-board and the ground recycling actions and the hazard wastes reusing. The hybrid strategy patterns were better ranked in general than on-board and the ground strategy patterns. This indicates in order to have a more powerful strategy pattern in reducing the wastes it is recommended to adopt the strategy that manage different sources of wastes (on-board, industrial and hazard).

The results of this chapter will benefit the academics and the industry experts alike. The academics have in-depth reporting for the relationship between the actions and the wastes indicators. Therefore, they can develop hypotheses for the future researchers. The experts in the industry have a clear idea about the different strategy patterns that could be adopted. In addition, the required actions and their impact in reducing the wastes. Despite the significant contribution of this chapter, some recommendations could be given for the future researchers. It is recommended to investigate a larger sample size to have a more valid results, it is recommended to investigate the interrelationship between the actions categories and the actions taken to have a better understanding for the independent, the moderating, the mediating and the dependent variables.

REFERENCES

American Plastics Council. (1997). *Stretch wrap recycling*. https://plastics.americanchemistry.com/Stretch-Wrap-Recycling/

Baxter, G., Srisaeng, P., & Wild, G. (2018). Sustainable Airport Waste Management: The Case of Kansai International Airport. *Recycling*, *3*(1), 6. doi:10.3390/recycling3010006

Blanca-Alcubilla, G., Bala, A., Hermira, J. I., De-Castro, N., Chavarri, R., Perales, R., ... Fullana-i-Palmer, P. (2018). Tackling International Airline Catering Waste Management: Life Zero Cabin Waste Project. State of The Art And First Steps. *Detritus*, *3*, 159.

GRI. (2017). *About GRI*. https:// www.globalreporting.org/Information/about-gri/Pages/default.aspx

Hondett, S. (2010). *Airline cabin waste recycling guide*. https://www.sustainableaviation.co.uk/wp-content/uploads/2018/06/Aircraft-Cabin-Waste-Recycling-Guide1.pdf

IATA. (2018). *Best industry practices for aircraft decommissioning (BIPAD)*. https://www.iata.org/publications/Documents/BIPAD.pdf

International Civil Air Transport Organization. (n.d.). *Waste management at Airports: ECO airport toolkit*.https://www.icao.int/environmentalprotection/Documents/Waste_Management_at_Airports_booklet.pdf

Li, X. D., Poon, C. S., Lee, S. C., Chung, S. S., & Luk, F. (2003). Waste reduction and recycling strategies for the in-flight services in the airline industry. *Resources, Conservation and Recycling*, *37*(2), 87–99. doi:10.1016/S0921-3449(02)00074-5

London Stansted Airport. (2010). *Making a material difference: London Stansted Airport waste management strategy 2010-2015*. https://live-webadmin-media.s3.amazonaws.com/media/3414/waste-strategy-stansted.pdf

Lorax, S. (2012). *The problem with pallets*. https://www.environmentalleader.com/2012/04/the-problem-with-pallets/

Mehta, P. (2015). Aviation waste management: An insight. *International Journal of Environmental Sciences*, *6*(1), 179.

Migdadi, Y. K. A. A. (2018). Identifying the best practices of airlines' green operations strategy: A cross-regional worldwide survey. *Environmental Quality Management*, *28*(1), 21–32. doi:10.1002/tqem.21575

Migdadi, Y. K. A. A. (2019). Identifying the effective taxonomies of airline green operations strategy. *Management of Environmental Quality*.

Migdadi, Y. K. A. A. (2020). Green Operational Strategy for Airlines: Content and Regional Analysis. In Cases on Green Energy and Sustainable Development (pp. 193-229). IGI Global.

Reiss, M., & Becker, A. (2018). *Managing cabin waste: market analysis and future concept. Airbus cabin waste management*. https://www.passengerexperienceconference.com/ RXUK/RXUK_PexConference/ Hamburg%202018/documents/ Airbus_Cabin_Waste_Solutions Presentation_ PEC_final_Handout. pdf?v=636609410168031406

Tofalli, N., Loizia, P., & Zorpas, A. A. (2018). Passengers waste production during flights. *Environmental Science and Pollution Research International*, *25*(36), 35764–35775. doi:10.100711356-017-0800-x PMID:29264848

Wieher, C. R. (2014). *Hazardous Waste Curriculum for Aviation Maintenance*. Brevard Teaching Research and Laboratories: Florida. https://www.broward.org/ EnvironmentAndGrowth/ EnvironmentalProgramsResources/ Publications/Documents/ CurriculumAviationMaint.pdf

Chapter 5
Airline Effective Water Consumption Strategies

ABSTRACT

This chapter reports the effective water consumption strategies. To achieve this aim, the water saving practices of 14 airlines from four regions were investigated. The data used in this chapter was a secondary published data in the annual sustainability reports. The data was retrieved from GRI (Global Reporting Initiatives) website. The special tailored data analysis techniques were used for this purpose. This chapter reported five effective strategic patterns. The strategic patterns were two facilities and buildings water saving, one facilities and buildings water recycling and recovery pattern, and two hybrid strategic patterns. This chapter helps the decision makers and the academics alike, since the impact of each action on reducing the water consumption under each action category of each pattern was reported.

INTRODUCTION

The total water consumption by the aviation industry was about 4,1016,913 m^3 (Airbus, 2018). Some airport made some actions to save water. The impact of these actions differs, Osaka's Kansai international Airport achieved -1.14 m3 water saving over the period 2002-2015 (Baxter, Erisaeng & Wild, 2018). Other airports as Frankfort AG reused about 100,000 m^3 a year. Orly and Charles de Gaulle Airport reused 70,000 m^3 of water a year and Atlanta airport reused 36.37 m^3 per month (Carvalho et al. 2013). Airlines took some actions

DOI: 10.4018/978-1-7998-4255-2.ch005

which achieved some progressing in term of the water saving, the achieved saving was on average -2.6%. The most effective cases achieved a saving by -8% (Migdadi, 2018), -3.75% and -14.16% (Migdadi, 2020).

The most of previous studies have investigated the water saving action of airports (e.g. Carvalho et al. 2013; Shamasky & Biochenko, 2016; Baxter, Srisaeng & Wild, 2019). The concern about reporting the effective water saving strategic patterns by airlines have not been the main research issue by the previous studies. This chapter bridged the gaps of previous studies by reporting the effective green operations strategies in saving the water. This chapter reported the practices of many airlines in different regions and developed strategic patterns. This chapter adopted the benchmarking methodology in making a comparison and reporting the most effective patterns. Accordingly, the main question of this chapter was: what were the most effective water saving strategic patterns that adopted by the airlines? To answer the question the following objectives were realized:

1. Ranking airlines according to its water consumption.
2. Identifying the effective water saving strategic patterns that adopted by airlines.
3. Identifying the significant green actions taken by the effective strategic patterns.
4. Identifying the impact of significant green actions on saving the water consumption.

This chapter is arranged in sections: the first one is the literature review, followed by the methodology and methods, then, the results of the empirical study, next, the discussion, and the final section is the conclusions.

LITERATURE REVIEW

This part of the chapter discussed the key literature of the airlines water saving. The first part of the literature review aimed to discuss the definition and the measures of water consumption indicator. The second part of the literature review discussed the green operations actions lists, categories, and the strategy patterns.

Airline Water Consumption Indicator

Water consumption is an indicator which measured as a cubic meter (m^3) of water consumed per passenger of the aircraft (Wyman, 2012), or the aircraft movement over a period of time (Baxter, Srisaeng & Wild, 2018). The large amounts of aviation water are consumed during the course of daily operations to support the operations such as the restrooms, the food service, the heating, the ventilating, and the air conditioning airfield services such as de-icing, construction, and firefighting. Also, maintenance activities such as the vehicle cleaning and the landscaping (Vurmaz & Boyacioglu, 2018). The realized saving of water consumption achieved by the airlines on average was -2.6%, the effective cases achieved a saving by -8% (Migdadi, 2018), -3.75% and -14.16% (Migdadi, 2020).

Airlines Water Saving Operational Actions and Strategy Patterns

The most of previous studies have investigated the water saving action of airports (e.g. Carvalho et al. 2013; Shamasky & Biochenko, 2016; Baxter, Srisaeng & Wild, 2019). Very limited studies have investigated the airlines practices (e.g. Migdadi, 2018; 2020). The majority of previous studies have reported the green practices to save the water consumption. The concern about reporting the effective water saving strategic patterns have not been the main research issue of the previous studies. The actions of airlines water saving could be classified into the recycling and recovery of maintenance water, the saving or maintenance water, the recycling and recovery of building and facilities water, and the saving of building and facilities water (Migdadi, 2918; 2020).

According to the limited studies of airlines, the significant adopted actions were the saving of maintenance water, with an impact of -4.95%, and the saving of water in facilities and buildings with an impact of -13.71% (Migdadi, 2020). Different initiatives have been adopted by the aviation sector to save water. For example, the grey water recycling. The grey water effluent from showers, kitchen, sinks, washing machines (Carvalho et al. 2013). The recycling of such kind of water could be done by using the anaerobic filter followed by UV disinfection (do Couto et al., 2015). In KIX airport, the percentage of recycled water was between 43% to 66.1% over the period 2010- 2015 (Baxter, Srisaeng & Wild, 2018). At Copenhagen airport two separate sever system to handle the wastewater and the rain water have been established. The first is the draining system to ensure that the airport area is not flooded, so the rain water is collected at the large rain water pools. The second system is the wastewaters treatment system, so the wastewaters are discharged to a treatment plant (Baxter, Srisaeng & Wild, 2019).

The treated or the recycled water is used for non-potable uses as toilet flashing, watering of gardens, and washing vehicles (Baxter, Srisaeng & Wild, 2018). However, the previous modern methods of water treatment require allocating of large areas for locating water treatment plants, as well as for arranging sanitary-protective zone around them (Shamasky & Biochenko, 2016). In addition, the using of information support system that helps in making the automated water saving process is devoted (Bieliatynsky et al, 2018). Other actions taken were controlling the water consumption through using meters to record the consumption of water. Also, water saving sanitary fixtures as flow limitation equipment on faucets, sensor operated wash basin faucets, taps with water saving nozzles, automatic ballcocks, replacing of water mains, and installation of float switch (Carvalho et al., 2013). Other practices could be adopted as the using of green leases by asking tenants of the airports to save water. For example, using incentives program, tenants sustainable rating, and tenant education (Hitz, 2015). Also training of the airport staff about the efficient water usages (Vurmaz & Boyocioglu, 2018).

METHODOLOGY AND METHODS

This part of the chapter discussed how the strategic patterns were identified, and how the effectiveness of these patterns were assessed. In this regards, the design of sample was discussed, also, how the data was collected was discussed. Then, how the data was analyzed to formulate the strategy patterns.

Sample Design

The sample of study was a convenient sample according to the availability of secondary published data. The water management practices were surveyed for 14 airlines. Table1 shows the distribution of sample between regions, it can be seen that the practices of six airlines from Asia, four airlines from Europe, two airlines from North America and two airlines from the Middle East were analyzed.

Table 1. Sample of study

Regions	Number of cases surveyed
A: Asia	6
E: Europe	4
NA: North America	2
ME: Middle East	2
Total	14

Sources of Data

The data was collected from the annual sustainability report of airlines. The reports were retrieved from Global Reporting Initiatives (GRI) database. The GRI is an independent international organization established in 1997 in the US. It seeks to help organizations to report their impact on sustainability. It develops standards of reporting sustainability, and publishes reports via its website (GRI, 2017).

Definition and Measures of Variables

The dependent variable was the amount of water consumed. However, the independent variables were the green water management actions. Consumed water was defined as m^3 of the used water by the airlines whether for the ground operations or during flights. However, the green water management actions were classified into different dimensions as the recycling and the recovery of maintenance water, the saving of maintenance water, the recycling and the recovery of water for facilities and buildings and the saving of water in the facilities and the buildings. The list of actions is presented in Table 2.

Data Analysis

How to assess the water consumption

Phase1: Computing the percentage of change in water consumption. The following steps were taken to achieve this phase:

1. Collecting the actual achieved performance in 2013 and 2016.
2. Compute the percentage of change by using the following formula:

Table 2. Airlines Water Management Actions

Action category	Actions list
Recycling and recovery of maintenance water	Recycling and reusing of waste maintenance water
	Refining of engine washing water
Saving maintenance water	Adopting "Eco Shine" method to clean the exterior of aircrafts using pads
Recycling and recovery of water for facilities and buildings	Using a rain water recovery system
	Recovering waste drinking water
	Recycling water used in the cooling towers of the building
Saving water in facilities and buildings	Optimization of swimming water
	Reducing the frequency of cleaning buildings
	Raising awareness of water saving among new employees
	Training of all cleaning staff in minimal water use
	Regular water tank maintenance
	Modification of water equipment
	Using minimal water for cooking and processing
	Adoption of water saving equipment
	Flow restriction and dual flash valves installed in all toilets
	Installing meters
	Eliminating least efficient dish washing machines
	Optimal use of water
	Launching water saving campaigns
	Identifying and resolving water leaks

(Source: Migdadi, 2018, 2020)

[water consumption in 2016- [water consumption in 2013] / [water consumption in 2013]X100%

Phase 2: Transform the percentage of change in water consumption into rating value as presented Table 3. The following steps were taken to achieve this phase:

1. Identifying the case achieved the best performance.
2. Dividing the performance achieved by each case by the best. The result is the relative performance.
3. Multiplying the relative performance by 5 (the maximum rating value of the scale). The result is the rating value of the performance.

Table 3. Rating Scale of Percentage of Change in Water Consumption

0	1	2	3	4	5
	Extremely low	Low	Moderate	High	Extremely high

Phase 3: Ranking the cases according to achieved rating value. The following steps were taken to achieve this phase:

1. Classifying the scale into three rating levels as following:
1. Step 1: identifying the turning value as following

[(5-0)]/3= 1.67

Step 2: identifying the levels by adding the turning value as following: From more than 0 to 1.67 is low level. From more than1.67 to 3.33 is moderate level. From more than 3.33to 5.00 is high level.

2. Assign the level for each case according to achieved rating value. The Assigned level represents the rank of the case performance.

How to identify the effective water saving strategic patterns

Phase 1: Computing the actions indices of water saving. The following steps were taken to achieve this phase:

1. Identifying the actions categories and actions lists. This study adopted the same classification of Migdadi (2018, 2020).
2. Computing the number of taken actions by each case under each category.
3. Dividing the number of taken actions by each case by the total number of actions could be taken. The result is a relative number called action adoption index.

Phase 2: Transforming the action taken index into a rating value as presented in the Table 4. The following steps were taken to achieve this phase:

1. Identifying the case achieved the best index value.
2. Dividing the action index achieved by each case by the best. The result is the relative action taken index.

Table 4. Rating Scale of Water Saving Actions Indexes

0	1	2	3	4	5
	Extremely low	Low	Moderate	High	Extremely high

3. Multiplying the relative index value by 5 (the maximum rating value of the scale). The result is the rating value of actions taken.

Phase 3: Identifying the relative level of action adoption according to achieved rating value

1. Classifying the scale into three rating levels as following:

1. Step 1: identifying the turning value as following:

[(5-0)]/3= 1.67

Step 2: Identifying the levels by adding the turning value as following: From more than 0 to 1.67 is low level. From more than1.67 to 3.33 is moderate level. From more than 3.33to 5.00 is high level.

2. Assign the level for each case according to the achieved rating value.

Phase 4: Summarizing the actions categories indices for all cases. The cases were ordered according to the ranking value of achieved water saving.

Phase 5: Identifying the actions categories with a moderate and a high rating value.

Phase 6: Identifying the cases took the shared actions categories with a moderate or a high level rating value.

Phase 7: Assigning the titles for the cases according to the most shared actions between the cases. These titles represent the different adopted patterns. If the patterns were adopted by the cases achieved an improvement in performance, then, the patterns would be recognized as the effective patterns.

How to compute the impact of actions taken by effective patterns on the achieved water saving

Phase 1: Identifying the actions taken by the higher effective cases in comparison with the lower effective and non-effective cases (which was adopted the same pattern). These actions were called the significant actions.
Phase 2: Counting the number of significant actions taken under each action category of the higher effective cases.
Phase 3: Computing the relative impact of each action category by dividing the number of significant action under each category by the total number of the significant actions.
Phase 4: Computing the differences in consumed water between the higher effective and lower effective or non-effective case as following:

Water consumption of the higher effective case- water consumption of the lower (or non-effective) case

Phase 5: Multiplying the relative impact of each action category by the differences in the water consumption between the higher and the lower effective or non-effective case. The result is the reduction in water consumption achieved by the action category.
Phase 6: Dividing the water saving achieved under the action category by the number of actions of the category. The result is the amount of water saving achieved by the action.
Phase 7: Computing the average water saving achieved by each action. The result is the average impact of the action under each category.

RESEARCH RESULTS

This part of this chapter presented the main results of the empirical study. The aim of this part was to identify the effective water saving strategies. The ranking of airlines according to the achieved water saving was presented first, then, the strategic patterns were presented as indicative models. The indicative models include the impact of actions adopted by each strategic pattern on the water saving. In order to facilitate the reading of this chapter results, Table 5 shows the acronyms of the strategy patterns alternatives.

Table 5. List of Acronyms Used in Presenting the Chapter Results

Acronym	Interpretation
FWR	Facilities and buildings water recycling and recovery
FWS	Facilities and buildings water saving
MWR	Maintenance water recycling and recovery
MWS	Maintenance water saving

Water Consumption of Airlines

Table 6 shows the water consumption of airlines. It can be seen that 80% (8 cases) saved the water consumption, and the remaining 20% increased the water consumption. The percentages of water saving was between -1.27% to -8.92%, the level of reduction according to the rating scale levels was high for three cases about 38% of the saved water cases. Also, three cases achieved a moderate saving in water consumption, however, two cases achieved a low saving in water consumption. The percentage of increase in water consumption was between 7.79% to 31.42%. The level of increase according to the rating scale level was high for the two cases.

Table 6. Water Consumption of Airlines

Case ID	% of change in water consumption	5-points rating of change in water consumption	Rating level
Reduced Water Consumption Cases			
ME1	-8.92%	5.00	+H
A4	-8.31%	4.66	+H
E4	-6.98%	3.92	+H
A5	-4.87%	2.73	+M
NA6	-4.13%	2.32	+M
E1	-3.75%	2.10	+M
E6	-2.00%	1.12	+L
A7	-1.27%	0.71	+L
Increased Water Consumption Cases			
NA5	7.79%	-4.37	-H
A2	31.42%	-5.00	-H

L: Low **M:** Moderate **H:** High
+: Improvement in the performance (Reduction in the water consumption)
-: Deterioration in the performance (Increase in the water consumption)

Airlines Effective Water Management Strategy Patterns

Table 7 shows the actions indices and the strategy patterns of water management. It can be seen that the levels of realized actions indices were low, moderate and high. For the purpose of this study, only the moderate and the high-level action indices were the significant taken actions by airlines. It can be seen that some cases took a combination of different significant actions as A4 and E1. Other cases took only one action as ME1, E4 and A7. On the other hand, the significant actions were taken by the reduced water consumption and the increased water consumption. The significant actions taken by the surveyed cases were used as a source to identify the strategic patterns of the cases. Accordingly, the last column of Table 7 shows these patterns. Each pattern has a title according to the most shared actions between the close patterns cases. The patterns adopted by airlines were facilities and buildings water saving strategy pattern, which was adopted by two cases of the reduced water consumption cases. Two cases adopted facilities and

Table 7. Actions Indices and Strategy Patterns of Water Management

level of % of change in water consumption	Case ID	Maintenance Water (MW)		Facilities and Buildings Water (FW)		Strategy Pattern	Strategy pattern title
		Recycling and recovery (R)	Saving (S)	Recycling and recovery (R)	Saving (S)		
Reduced Water Consumption Cases							
+H	ME1	0.00	0.00	0.00	3.70 (H)	FWS	Facilities and Buildings Water Saving
+H	A4	5.00 (H)	0.00	5.00 (H)	5.00 (H)	MWR-FWR-FWS	Hybrid
+M	E4	0.00	0.00	0.00	3.70 (H)	FWS	Facilities and Buildings Water Saving
+M	A5	0.00	0.00	0.00	1.30 (L)	----------	----------
+M	NA6	0.00	0.00	0.00	1.30 (L)	----------	----------
+L	E1	2.50 (M)	5.00 (H)	0.00	2.41 (M)	MWR-MWS-FWS	Hybrid
+L	E6	0.00	0.00	0.00	1.30 (L)	----------	----------
+L	A7	0.00	0.00	5.00 (H)	1.30 (L)	FWR	Facilities and Buildings Water Recovery
Increased Water Consumption Cases							
-H	NA5	0.00	0.00	2.46 (M)	0.00	FWR	Facilities and Buildings Water Recovery
-H	A2	0.00	0.00	0.00	0.00	----------	----------

(L): Low (M): Moderate (H): High

buildings water recovery pattern. One of them was adopted by the reduced water consumption cases. Hybrid strategy pattern, which was adopted by two cases, all of them were of the reduced water consumption cases.

Facilities and Buildings Water Saving Strategic Patterns

Table 8 shows the effective water saving strategic patterns. It can be seen that two patterns were adopted. The first one was (FWS), which was adopted by ME1. The second pattern was (FWS), which was adopted by E4. The first pattern was the most effective, the achieved water saving was -5.30%. The second pattern achieved a saving by -3.61%. Despite the sharing of the actions categories between the two strategic patterns, but the actions taken was completely different. The first pattern adopted three actions, which were the raising awareness of water saving among new employees, the training of all cleaning staff in the minimal water use, and modifying the water equipment. The saving in water achieved by these three action was the same, it was

Table 8. Effective Facilities and Buildings Water Saving Strategic Patterns

Case ID	Strategy Pattern	Actions category Total Impact	Significant actions taken	Average Relative % of change in Water Consumption
ME1	FWS	**Facilities and Buildings Water Saving (FWS)** Total Impact= -5.30%	Raising awareness of water saving among new employees	-1.77%
			Training of all cleaning staff in minimal water use	-1.77%
			Modification of water equipment	-1.77%
Total impact of the pattern				-5.30%
E4	FWS	**Facilities and Buildings Water Saving (FWS)** Total Impact= -3.61%	Identifying and resolving water leaks	-1.39%
			Installing meters	-0.83%
			Optimal use of water	-1.39%
Total impact of the pattern				-3.61%

-1.77%. In addition, these three actions were more effective than all actions adopted by the second pattern. The second pattern, which was adopted by E4 contained three actions. These actions were identifying and resolving the water leaks, installing the meters, and the optimal use of water. The least effective action taken by the second pattern was installing the meter. The achieved saving in water by this action was -0.83%, and these are the least effective action taken by both strategic pattern. The impact of the first and the third action taken by E4 was the same, which was -1.39%.

Effective Facilities and Buildings Water Recovery Strategic Patterns

Table 9 shows the strategic pattern of facilities and buildings water Recovery and recycling (FWR). This is the only effective pattern, which was adopted by A7. The total water saving achieved was -22%. Two taken actions, which were using a rainwater recovery system, and the recycling of water used in the cooling towers of the building. The impact of these two action in saving water was the same, it was -11%.

Table 9. Effective Facilities and Buildings Recovery Strategic Patterns

Case ID	Strategy pattern	Action category Total Impact	Significant actions taken	Average Relative % of change in Water Consumption
A7	FWR	**Facilities and Buildings Water Recovery and Recycling (FWR)** Total Impact= -22%	Using a rain water recovery system	-11%
			Recycling water used in the cooling towers of the building	-11%
Total impact of the pattern				-22%

Effective Hybrid Water Management Strategic Patterns

Table 10 shows the effective hybrid water management strategic patterns. Two patterns were adopted. The first pattern was adopted by the case A4. This pattern includes the actions categories of the maintenance water recycling and recovery, the facilities and the buildings water recycling and recovery, and

Table 10. Effective Hybrid Strategic Patterns

Case ID	Strategy pattern	Action category Total Impact	Significant actions taken	Average Relative % of change in water consumption
A4	MWR-FWR-FWS	**Maintenance Water Recycling Recovery (MWR)** Total Impact= -4.06%	Recycling and reusing of waste maintenance water	-0.77%
			Refining of engine washing water	-3.29%
		Facilities and Buildings Water Recycling Recovery (FWR) Total Impact= -4.83%	Using a rain water recovery system	-2.42%
			Recovering waste drinking water	-2.42%
		Facilities and Buildings Water Saving (FWS) Total Impact= -9.66%	Optimization of swimming water	-2.42%
			Reducing the frequency of cleaning buildings	-2.42%
			Adoption of water saving equipment	-2.42%
			Optimal use of water	-2.42%
Total impact of the pattern				-18.55%
E1	MWR-MWS-FWS	**Maintenance Water Recycling Recovery (MWR)** Total Impact= -4.63%	Refining of engine washing water	-4.63%
		Maintenance Water Saving (MWS) Total Impact= -4.63%	Adopting "Eco Shine" method to clean the exterior of aircrafts using pads	-4.63%
		Facilities and Buildings Water Saving (FWS) Total Impact= -9.26%	Installing meters	-4.63%
			Eliminating least efficient dish washing machines	-4.63%
Total impact of the pattern				-18.50%

the facilities and the buildings water saving. The second pattern, which was adopted by the case E1. The second pattern includes the actions categories of the maintenance water recycling and recovery, the maintenance water saving, and the facilities and the buildings water saving. The two strategic patterns were very close in term of the total water saving. It was -18.55% achieved saving by the first strategy pattern, and it was -18.50% achieved saving by the second strategy pattern. Two shared actions categories between the two strategic patterns, which were the maintenance water recycling and recovery, and the facilities and the buildings water saving.

The shared maintenance water recycling and recovery action between the two strategic patterns was the refining of engine washing water. The saved water was -3.29% by the first pattern, and by -4.63% by the second strategy pattern. The impact of all remaining actions that adopted by the first pattern was equal, the achieved saving was -2.42%. The least effective action adopted by the first pattern was recycling and reusing of the waste maintenance water, the water saving achieved by this action was -0.77%. All actions adopted by the second pattern achieved the same saving of the consumed water, the saving achieved by each action was -4.63%.

Summary of Effective Water Consumption Strategic Patterns

Table 11 shows a summary of the effective water consumption strategic patterns. It can be seen that three strategy patterns achieved a high saving of the consumed water. These patterns were facilities and buildings water recycling and recovery (FWR), hybrid (MWR-MWS-FWS) and hybrid (MWR-FWR-FWS). The saving achieved by these three patterns was -22%, -18.55% and -18.50% respectively. Two strategic patterns achieved a low saving in the consumed water, these pattern were facilities and buildings water saving (FWS), which was adopted by ME1, and facilities and buildings water saving (FWS), which was adopted by E4. The saving achieved by these two patterns was -5.30% and -3.61% respectively.

Table 11. Summary of Airline Effective Airline Water Consumption Strategy Patterns

Case ID	Strategy pattern title	Strategy pattern	Relative average % of change in water consumption	Maintenance water Recycling and recovery (MWR)	Maintenance water Saving (MWS)	Facilities and buildings water Recycling and recovery (FWR)	Facilities and buildings Water Saving (FWS)
A7	Facilities and Buildings Water Recycling and Recovery	FWR	-22% (H)	---------	---------	-22%	--------
E1	Hybrid	MWR-MWS-FWS	-18.55% (H)	-4.63%	-4.63%	--------	-9.26%
A4	Hybrid	MWR-FWR-FWS	-18.50% (H)	-4.06%	--------	-4.83%	-9.66%
ME1	Facilities and Buildings Waster Saving	FWS	-5.30% (L)	---------	---------	---------	-5.30%
E4	Facilities and Buildings Water Saving	FWS	-3.61% (L)	---------	---------	---------	-3.61%

(L): Low **(M)**: Moderate **(H)**: High

DISCUSSION

The most effective strategy pattern of (FWS) was that adopted by ME1. According to this pattern, three effective actions were taken. The impact of these actions were equal, which was -1.77%, and these actions were more effective than those adopted by the second pattern that was adopted by E4. The actions adopted by ME1 were raising the awareness of water saving among the new employees, the training of all cleaning staff on the minimal water use, and the modification of water equipment. This issue has been investigated in other industries as hotels. The results of the hospitality previous studies found that the environmental attitudes have an effect on the environmental behavior of both the home and the workplace behaviors (Wells, Taheri, Gregory-Smith & Manika, 2016). In addition, the employees' environmental knowledge determines the environmental awareness and concern. These are the significant factors in determining the employees' environmental behavior and the green practices. Accordingly, the environmental performance of employees is closely related to the green management factors as the selecting of people with a good environmental knowledge, and providing a continuous environmental training to the staff (Chan, Hon, Okumus & Chan, 2017).

The effective actions taken by E4 were the resolving of water leakage, installing the meters, and the optimal use of water. The water saving achieved by of these actions was -1.39%, -0.83% and -1.39% respectively. This confirms the previous studies results as Carvalho et al. (2013). The detection and localization of leaks constitute a priority in the management of drinking water systems in general, the application of the improved method allowed the detection of 25 unreported leaks and decreased the Non-Revenue Water (NRW) level by 36% (Farah & Shahrour, 2017). Accordingly, the leak detection analytics and alerts will minimize the water loss and the damage due to the leaks (Schultz, Javey & Sorokina, 2018). The using of water meter will help in protecting the environment by conserving water, you will become more aware of your water use when you become metered as the more you use, the more you pay (Save Water Save money, 2019), the metered customers usually use 15% to 20% less water than the un-metered customers (Alliance for Water Efficiency, 2019).

The two effective facilities and buildings water recovery strategy actions were equal in their water saving, which was -11% each. These actions were using a rainwater recovery system, and the recycling of water used in the cooling towers of the buildings. These actions were adopted widely by the airports as reported by the previous studies (e.g. do Couto et al., 2015; Shamasky & Biochenko, 2016; Baxter, Srisaeng & Wild, 2018). Gray water recycling and economical use of the rainwater can be a valuable alternative source of water, especially for the non-potable uses (Stec & Kordana, 2015). Cooling water system is used widely in the industry, where large amounts of fresh water consumed each year by the cooling towers (Ma et al., 2018).

The two hybrid strategies were very close in term of the total water saving. The shared actions between the two patterns were refining the engine washing water. However, the impact of this action as adopted by the second pattern (which was adopted by E1) was more effective than the first pattern (which was adopted by A4). This could be related to the kind of washing technology used. Each technology consumed a different amount of water. For example, the traditional engine washing by flushing water in the engine, consumes large amount of water, and the waste water generated requires more actions in the recycling later. However, adopt more advanced washing technologies reduce the amount of water use since it adopts a dual nozzle 360 arrangement, also it sprays pre-heated water into the core engine (Premagard Aviation, 2019). Adopting "Eco Shine" method to clean the exterior of aircrafts by using pads was one of the effective actions adopted by the second strategy pattern (which was adopted by E1). This kind of washing is a dry washing of the aircraft body, this kind of washing save more than 95% of water used in the non-dry washing, and more time saving (Aviostar, 2019). Eliminating least efficient dish washing machines was adopted by the second strategy pattern. This action is rational since the new dishwashers save about 3870 gallons of water over its lifecycle (Energystar, 2019).

CONCLUSION

The aim of this chapter was to report the effective airlines water saving strategies. This research issue has been raised by limited studies. The most of previous studies have reported water saving green practices of the airports. The main results of this study were about 80% of the airlines achieved the saving of water. Different water saving strategy patterns were adopted. The effective strategy patterns were two facilities and buildings water saving strategy pattern, two hybrid strategy patterns and one facilities and building water recovery strategy patterns.

The results of this chapter will benefit the academics and the industry experts alike. The academics have in-depth reporting for the relationship between the actions and the green operations actions. Therefore, they can develop hypotheses for the future researchers. The experts in the industry have a clear idea about the different strategy patterns that could be adopted. In addition, the required actions and their impact on reducing the water consumption. Despite the significant contribution of this chapter, some recommendations could be given for the future researchers. It is recommended to investigate a larger sample size to have more valid results, it is recommended to investigate the interrelationship between the actions categories and the actions taken to have a better understanding of the independent, the moderating, the mediating and the dependent variables.

REFERENCES

Airbus. (2018). *Annual report 2018*. https://www.airbus.com/content/dam/corporate-topics/facts-and-figures/annual-report/Information-document-2018-pages-52-to-57.pdf

Alliance for Water Efficiency. (2019). *Submetering*. https://www.allianceforwaterefficiency.org/resources/topic/submetering

Aviostar. (2019). *Dry wash.* https://www.avio-star.com/dry-wash

Baxter, G., Srisaeng, P., & Wild, G. (2018). An Assessment of Sustainable Airport Water Management: The Case of Osaka's Kansai International Airport. *Infrastructures, 3*(4), 54. doi:10.3390/infrastructures3040054

Baxter, G., Srisaeng, P., & Wild, G. (2019). An Assessment of Airport Sustainability: Part 3—Water Management at Copenhagen Airport. *Resources, 8*(3), 135. doi:10.3390/resources8030135

Bieliatynskyi, A., Osipa, L., Yaroshchuk, L., & Cheropkin, E. (2018). *Information support of automated control subsystem of the airport's water-saving processes.* 21 conference of junior researchers science. Future of Lithuania, Vilnius.

Carvalho, I. D. C., Calijuri, M. L., Assemany, P. P., Silva, M. D. F. M., Moreira Neto, R. F., Santiago, A. D. F., & Souza, M. H. B. D. (2013). *Sustainable airport environments: a review of water conservation practices in airports.* Academic Press.

Chan, E. S., Hon, A. H., Okumus, F., & Chan, W. (2017). An empirical study of environmental practices and employee ecological behavior in the hotel industry. *Journal of Hospitality & Tourism Research (Washington, D.C.), 41*(5), 585–608. doi:10.1177/1096348014550873

do Couto, E. D. A., Calijuri, M. L., Assemany, P. P., da Fonseca Santiago, A., & Lopes, L. S. (2015). Greywater treatment in airports using anaerobic filter followed by UV disinfection: An efficient and low cost alternative. *Journal of Cleaner Production, 106*, 372–379. doi:10.1016/j.jclepro.2014.07.065

Energystar. (2019). *Dishwashers.* https://www.energystar.gov/products/appliances/dishwashers

Farah, E., & Shahrour, I. (2017). Leakage detection using smart water system: Combination of water balance and automated minimum night flow. *Water Resources Management, 31*(15), 4821–4833. doi:10.100711269-017-1780-9

GRI. (2017). *About GRI.* https:// www.globalreporting.org/Information/about-gri/Pages/default.aspx

Hitz, C. (2015). *Water conservation at Vancouver International Airport: strategies for water conservation and efficiency*. University of British Columbia.

Ma, J., Irfan, H. M., Wang, Y., Feng, X., & Xu, D. (2018). Recovering Wastewater in a Cooling Water System with Thermal Membrane Distillation. *Industrial & Engineering Chemistry Research*, *57*(31), 10491–10499. doi:10.1021/acs.iecr.8b00317

Migdadi, Y. K. A. A. (2018). Identifying the best practices of airlines' green operations strategy: A cross-regional worldwide survey. *Environmental Quality Management*, *28*(1), 21–32. doi:10.1002/tqem.21575

Migdadi, Y. K. A. A. (2020). Green Operational Strategy for Airlines: Content and Regional Analysis. In Cases on Green Energy and Sustainable Development (pp. 193-229). IGI Global.

Premagard Aviation. (2019). *Cyclean engine washing*. http://www.permagard.com.au/divisions/aviation/products/cyclean.html

Save Water Save Money. (2019). *Can I save water by installing meters*. https://www.savewatersavemoney.co.uk/water-efficiency-tips-advice/view/109/all-you-need-to-know-about-water-meters.html

Schultz, W., Javey, S., & Sorokina, A. (2018). Smart Water Meters and Data Analytics Decrease Wasted Water Due to Leaks. *Journal - American Water Works Association*, *110*(11), 24–30. doi:10.1002/awwa.1124

Shamanskyi, S., & Boichenko, S. (2016). Development of environmentally safe technological water disposal scheme of aviation enterprise. *Восточно-европейский журнал передовых технологий*, *6*(10), 49-57.

Stec, A., & Kordana, S. (2015). Analysis of profitability of rainwater harvesting, gray water recycling and drain water heat recovery systems. *Resources, Conservation and Recycling*, *105*, 84–94. doi:10.1016/j.resconrec.2015.10.006

Vurmaz, M. Ö., & Boyacioglu, H. (2018). Airport Water Consumption Foot printing. *Environment and Ecology Research*, *6*(6), 519–524. doi:10.13189/eer.2018.060601

Wells, V. K., Taheri, B., Gregory-Smith, D., & Manika, D. (2016). The role of generativity and attitudes on employees home and workplace water and energy saving behaviours. *Tourism Management*, *56*, 63–74. doi:10.1016/j.tourman.2016.03.027

Wyman, O. (2012). *Guide to airport performance measures. Airports Council international*. ACI.

Chapter 6
Airline Effective Noise Management Strategies

ABSTRACT

This chapter reports the effective noise management strategies of the airlines. To achieve this aim, the practices of 18 airlines from five regions were investigated. The data used in this chapter was secondary published data in the annual sustainability reports. The data was retrieved from GRI (Global Reporting Initiatives) website. The special tailored data analysis techniques were used for this purpose. The chapter reported 17 effective strategic patterns. The strategic patterns were seven hybrid patterns, four conformance with ICAO standards patterns, two flights operating patterns, one conformance with FAA standards pattern, and three aircraft design patterns. This chapter helps the decision makers and the academics alike.

INTRODUCTION

The concern about the aviation noise tracking and management date back to 1960s. Over this period the aircrafts engine technology has became more common place, the international aviation became more popular in Europe and USA, more advancement in the wide-bodied, jet engine aircrafts, the dawning of the jet age, and more protesting from the local community around airports (Civil Aviation Authority, 2014). Noise effects could be behavioural or physiological. The behavioural impacts are those interfere the normal living as speech interferences, sleep interferences and disruption of the work activity.

DOI: 10.4018/978-1-7998-4255-2.ch006

However, the physiological or the medical impacts are those related to the chronical effects as sever annoyance that cause the hearing loss (Civil Aviation Authority, 2014; Glasgow Airport, 2014). Despite the average level of noise is lower than before and lower than the road and the rail transportation noise, the frequency of aircrafts movements and hence noise event has generally increased (Glasgow Airport, 2014), and tends to cover larger geographical areas and is more difficult to mitigate when its compared to the road and the rail noise (Civil Aviation Authority, 2014).

Many studies have investigated the noise, some of them have investigated the abatement operations procedure (e.g. Fillipone, 2017; Gagliardi Teti & Licitra, 2018; Pretto, Giannattas, De Gennavo & Zanon, 2019). Some studies have investigated the reduction of noise at source (e.g. Torija et al., 2019) and other studies have investigated the land-use planning (e.g. Flores, Asensio, Gagliardig & Licitra, 2019). Very limited studies have investigated the airlines green operations strategies (e.g. Migdadi, 2018, 2019, 2020); these studies have investigated the general strategic patterns. The strategic patterns of airlines noise management have not been investigated by the previous studies.

This chapter bridged the gap of previous studies by reporting the airlines noise reduction strategies. This study reported the practices of many airlines in different regions and developed the strategic patterns. This chapter adopted the benchmarking methodology by comparing the practices of cases with the best practices case to report the most effective patterns. Accordingly, the main question of this chapter was: what were the effective green strategic patterns that adopted by airlines to reduce noise?, to answer the question the following objectives were realized:

1. Identifying the effective noise management strategic patterns that adopted by airlines.
2. Identifying the significant green actions that taken by the strategic patterns.

This chapter is arranged in sections: the first one is the literature review, followed by the methodology and methods, then, the results of the empirical study, next, the discussion, and the final section is the conclusions.

LITERATURE REVIEW

This part of the chapter discussed the key literature of the airlines noise management. The first part of the literature aimed to discuss how the airline noise is measured, so different methodologies and models were presented, also the factors and assumptions that should be considered in measuring airline noise were discussed. The second part of the literature review discussed the green operations actions lists, categories, and the strategy patterns.

Airline Noise Measurement

Noise is the unwanted level of sound (Civil Air Navigation Service Organization, 2015). The aviation noise is caused by air passing over the aircraft body, which causes turbulence and friction, or the noise is caused by the sounds of engine moving parts, and the sound of air being expelled at a high speed (Glasgow Airport, 2014). UK government assumes that the community will be annoyed by aircraft noise above 57 dBLAeq (Glasgow Airport, 2014). The noise effects could be measured objectively and quantitatively as the noise exposure indicators, or subjectively and qualitative as the behavioral indicators, whatever the approach adopted, the noise index should be simple, accurate and practical (Civil Aviation Authority, 2014).

The aviation noise metrics could be classified into the single event based metrics, or the individual aircraft noise over a short period of time. Other metric is the time based noise metrics, or the average noise over a defined period. Threshold based metrics, which is the total number of time a defined noise threshold is exceeded in a specific period as 24 hours. Track based metrics, which is the tracking of density, respite and the number periods of flight noise. Different indicators could be adopted to measure each kind of the metrics. For example, the maximum sound level generated during event (L_{MAX}) is the widely adopted single event indicator. Weighted average sound level at a location over a period of time (LAeq) is one of the time-based indicators (CANSO, 2015).

Different indicators could be adopted by the national or the regional civil air authorities. For example, in the UK government, equivalent continuous sound level (dBLAeq) indicator is used. Accordingly, the noise is measured over the busiest 16 hours of the day, between 7am. to 23:00 over the busiest three months of the year, form mid Jun to mid-September (Glasgow Airport, 2014). The widely adopted indicator by ICAO is EPNdB, which is the effective perceived noise per decibels. This indicator measures the relative noisiness of an individual aircraft pass-by-event. This indicator is used by ICAO for the aircraft's noise certification under ICAO Annex 16. Chapter 2, 3, 4, and 14 for the individual aircraft, and not for the noise exposure by the airport. The index adopted by ICAO placing aircrafts into six categories on noise performance, ranging from A to F. According to Ch.3 rating scale: (A) rating reflects at least 20 or more reduction in EPNdB, (B) rating reflects 15 or more reduction, (C) reflects 10 or more reduction, (D) is 5 or more reduction, (E) 0 or more reduction, and (F) less than 0 (Airport Council International, 2010).

Airlines Noise Management Actions and Strategy Patterns

The most famous approach of managing noise is that developed by ICAO, and which is called "Balanced Approach to Aircraft Noise Management". According to these polices, the noise of each airport will be measured and analyzed according to the reduction of noise through tracking four measure dimensions: reduction of noise at source, land-use planning and management, noise abatement operations procedures and operations restrictions. These policies are published in ICAO Doc. 9829, "Guidance on the Balanced Approach to Aircraft Noise Management" (ICAO, 2019a).

Reduction of noise at source means the reduction of noise generated by the aircrafts, so the manufacturers should improve the aircraft and engine design to have lower noise aircrafts, and the airlines should replace their fleets with lower noise aircrafts. This measure of noise has been controlled since 1970s by developing limits for aircraft noise in a form of "Standards and Recommended Practices" (SARPs) constrained in Annex 16, Ch.2 Vol.1, which became applicate in 1972. According to these standards, the noise is a function of "Maximum Take-off Mass" (MTOM). Accordingly, the heavier aircrafts with greater transport capacity produces more noise than the lighter aircrafts. These standards were revised and improved more than one time to cope with the advancement of aircrafts technologies as engines. Accordingly, more stringent noise standard were developed by ICAO in 1977, this was called Ch.3. Further stringent standards were developed in 2001, which was called Ch.4. In 2014 ICAO council adopted the new Ch.14 noise standards (ICAO, 2019b). ICAO's Committee on Aviation Environmental Protection (CAEP) is the competent ICAO's body for developing and revising Annex 16 aircraft noise and the engine emissions (Airport Council International, 2010).

According to the land-use planning management, the noise sensitive areas, like the residences, the hospitals and the schools around the airports, should be avoided as much as possible from the aircraft noise by zoning the land-use around the airports. Local municipal governments and airports are usually responsible for land-zoning. For example, the home in very high noise areas could be purchased by the airports (CANSO, 2015). The guidelines of this subject is contained in Annex 16, Vol.1, part IV, and ICAO Doc. 9184. Airport Planning Manual, part 2-land use and environmental control (ICAO, 2019c). The third dimension of the ICAO balanced approach is the noise abatement operational procedures. According to this dimension, the daily operating of aircrafts as the preferential runway, the routes, the noise abatement procedures for take-off and landing will affect the noise. Several documents were developed by ICAO for this purpose as Doc. 8168 (procedures for air navigation service-Aircraft operations) (PANS-OPS) part I, part II. Doc. 9931-Continuous Descent Operations (CDO) manual, Doc. 9993-Continuous Climb Operations (CCO) manual, Doc. 9888- Review of noise abatement research and development and improvement and implementation project (ICAO, 2019d).

The operating restrictions are restrictions on the operating of airports. The restrictions include curfews, night time restrictions, noise quotes/budgets and cap rules. These restrictions are described in Ch.7 of ICAO Doc. 9829 (ICAO, 2019e). IATA has many concerns about this dimension of ICAO's " Balanced Approach to Aircraft Noise Management". According to IATA, the operations restrictions should be applied on the airports demonstrate noise problems. The operational restrictions should be applied after a full assessment for the benefits gained from other measures of the balanced approach. All relevant stakeholders as airlines should be involved in the decision of operating restrictions before making the decisions. Operating restrictions should be according to a particular level of certifications and not according to the type of operations, the airlines should be noticed before a sufficient time and should be made gradually (IATA, 2013). The night flight restrictions has many negative consequences on the airline industry. Night flights are critical for cargo and express operations, such restriction will delay the delivery of products, which will delay the products delivery over the supply chain. Moreover, the rescheduling of night flights will make the day flight schedules more crowed (IATA, 2018).

Many studies have investigated the noise. Some of them have investigated the abatement operations procedure (e.g. Fillipone, 2017; Gagliardi Teti & Licitra, 2018; Pretto, Giannattas, De Gennavo & Zanon, 2019). Some studies have investigated the reduction of noise at source (e.g. Torija et al., 2019), and other studies have investigated the land-use planning (e.g. Flores, Asensio, Gagliardig & Licitra, 2019). Very limited studies have investigated the airlines green operations strategies (e.g. Migdadi, 2018, 2019, 2020). These studies have investigated the general strategic patterns. The strategic patterns of airlines noise management have not been investigated by previous studies.

Today aircrafts are much quieter than that manufactured before 20, 30 or 40 years ago. For example, the noise footprint of B787 Dreamliner is 60% smaller than similarly sized aircrafts (Glasgow Airport, 2014), also the last version of B747, B747-8 intercontinental claim a 30% noise performance improvement over that of its predecessor B747-400, and A380 is one quietest wide-body jet aircraft (Civil Aviation Authority, 2014). A380 powered by Rolls-Royce Trent 900, and B787 powered by Rolls-Royce Trent 1000 are more quiet than the predecessors are. The engines of these aircrafts increased engine bypass, ratio, and nacelles with a zero-splice intake liners (Sustainable Aviation, 2018). Auras tonality of different engines of the aircrafts is found to be the main factor of perceived noise (Torija et al., 2019). The noise is affected by the flight events (departure and arrival) and the reconstruct of aircraft flight path (Pretto, Giannattas, De Gennavo & Zanon, 2019). The noise reduction of -1 EPNL is possible with a 450m landing and within a 3.5-degree glide slope (Fillppone, 2017). In addition, scheduling the landing of a sequence of aircrafts will minimize the noise (Rodrignes-Diaz, Adenso-Diaz & Gonzalez-Torre, 2019). Moreover, the actual take-off weight and both the aircraft ground speed and altitude are the most significant factors affected the noise emissions level (Gagliardi, Teti & Licitra, 2018).

The interacting of buildings with sound waves will modify the sound pressure that produced by the aircraft noise (Flores, Asensio, Gagliardig & Licitra, 2019). London residents are more suffering from the severity of Heathrow Airport noise. Accordingly, more actions should be taken to reduce the noise as the flight paths should be redesigned, the night flights should be minimized (London Assembly Environment Committee, 2019). There are noise trackers systems at major airports such as Heathrow, these systems gather information on the noise that made by landing or departing aircrafts. The guidance for controlling noise at Heathrow airport is Environmental Protection Act 1990, section 79 (6), and directive 2002/30/EC on the Introduction of Noise-related Operating Restriction at Community Airports (Butcher, 2017). Federal Aviation Agency adopted an integrative approach to evaluate the aircraft noise during the defined operations around the airport, the guidance to this is ICAO Doc. 9911 and Aviation Environment Design Tool (AEDT) (CANSO, 2015).

METHODOLOGY AND METHODS

This part of the chapter discussed how the strategic patterns were identified. However, the effectiveness of strategy patterns were not analyzed, since the majority of airlines are not reporting the noise performance. In this regards, the design of sample was discussed, also, how the data was collected was discussed too. Then, how the data was analyzed to formulate the strategy patterns was discussed in details.

Sample Design

The sample of this study was a convenient sample according to the availability of secondary published data. The noise operations were surveyed for 18 airlines. Table1 shows the distribution of sample between regions, it can be seen that the noise operations management was surveyed for five airlines in Aisa and Europe. In addition, four airlines were surveyed in North America, three airlines in the Middle East, and one airline in South America.

Table 1. Sample of study

Regions	Number of airlines
A: Asia	5
E: Europe	5
NA: North America	4
ME: Middle East	3
SA: South America	1
Total	18

Sources of Data

The data was collected from the annual sustainability report of airlines. The reports were retrieved from Global Reporting Initiatives (GRI) database. Most of these reports were retrieved from the GRI. The GRI is an independent international organization established in 1997 in the US. It seeks to help organizations to report their impact on sustainability. It develops standards of reporting the sustainability, and publishes the reports via its website (GRI, 2017).

Definition and Measures of Variables

The noise indicator has not surveyed in this study since the most of airlines are not reporting the noise performance. Accordingly, no independent and dependent variables in this chapter. The only investigated variable by this study was the noise management actions. These actions were classified into different dimensions as the aircraft design, the aircraft conformance with ICAO standards, the aircraft conformance with FAA standards, the conformance with airport fly quiet program, and the flight operating. Each dimension of actions was measured by set of actions. The list of actions is presented in Table 2.

Data Analysis

The aim of this part was to discuss how to identify the noise management strategic patterns. Accordingly, the following phases were followed;

Phase 1: Computing the actions indices of noise. The following steps were taken to achieve this phase:

1. Identifying the actions categories and actions lists by reviewing the related literature.
2. Computing the number of actions taken by each case under each category.
3. Dividing the number of taken actions by each case by the total number of actions that could be taken. The result is a relative number called action adoption index.

Phase 2: Transforming the action taken index into rating value as presented in Table 3. The following steps were taken to achieve this phase:

1. Identifying the case achieved the best index value.
2. Dividing the action index achieved by each case by the best. The result is the relative action taken index.
3. Multiplying the relative index value by 5 (the maximum rating value of the scale). The result is the rating value of the actions taken.

Table 2. Airlines Noise Management Actions

Actions Dimensions	Actions
Aircraft Design	Fleet renewal (B737, B737 MAX, A320, A350, B787)
	Modification of aircraft design to lower noise ambient as winglets design
Aircraft conformance with ICAO noise Standards	% of aircrafts conforms with ICAO's Ch.2
	% of aircrafts conforms with ICAO's Ch.3
	% of aircrafts conforms with ICAO's Ch.4
	% of aircrafts conforms with ICAO's Ch.14
Aircraft conformance with FAA standards	% of aircrafts meets stage 4 criteria of FAA
Conformance with Airport fly quiet program	The rank according to noise performance (short haul fleet) of Heathrow fly quiet program
	The rank according to noise performance (long haul fleet) of Heathrow fly quiet program
Flights operating	Noise hearing screening program
	Night flight rescheduling and procedures improvement according to CAD (Civil Aviation Department) procedures
	Day flight procedures improvement according to CAD procedures
	Improve departure and approach procedures with civil aviation authority abatement procedures
	Electrical taxiing
	Area navigation (RNAV) departure
	Performed well with respect to continuous descent approach (CDA)
	Controlling and monitoring noise from airports
	Optimum flights routes
	Rescheduling takeoffs and landing at less undesirable time from a noise perspective
	Reduce using of APU
	Practicing noise abatement procedures during takeoff
	Adopting continuous climb operation (CCO)
	Practicing noise abatement procedures during landing (reduce flap, delayed flap and idle reserve)

Phase 3: Identifying the relative level of action adoption according to achieved rating value. The following steps were taken to achieve this phase:

1. Classifying the scale into three rating levels as following:

   Step 1: identifying the turning value as following

[(5-0)]/3= 1.67

Table 3. Rating Scale of Noise Actin Indexes

0	1	2	3	4	5
	Extremely low	Low	Moderate	High	Extremely high

Step 2: Identifying the levels by adding the turning value as following

From more than 0 to 1.67 is low level. From more than1.67 to 3.33 is moderate level. From more than 3.33to 5.00 is high level.

2. Assign the level for each case according to the achieved turning value.

Phase 4: Identifying the actions categories with a moderate and a high level rating value.

Phase 5: Identifying the cases took shared actions categories with a moderate or a high level rating value.

Phase 6: Assigning the titles for the cases according to most shared actions between cases.

RESEARCH RESULTS

This part of this chapter presented the main results of the empirical study. The aim of this part was to identify the noise management strategies. Accordingly, the first part presented the summary of strategy patterns, followed by the detail discussion for the taken actions by each strategy patterns. In order to facilitate the reading of this chapter results Table 4 shows the acronyms used in presenting the chapter results.

Airlines Noise Management Strategic Patterns

Table 5 shows the actions indices and the strategy patterns of noise management. It can be seen that the levels of realized actions indices were low, moderate and high. For the purpose of this study, only the moderate and the high-level actions indices were recognized as the significant taken actions by the airlines. It can be seen that some cases took a combination of different significant actions, as E2, E3 and E1. Other cases took only one action as A6 and NA3. The significant actions taken by the surveyed cases were used as a source

Table 4. List of Acronyms Used in Presenting the Chapter Results

Acronym	Interpretation
AS	Conformance with Airport fly quiet program
D	Aircraft Design
FAAS	Aircraft conformance with FAA noise standards
FO	Flights operating
ICAOS	Aircraft conformance with ICAO Noise Standards

to identify the strategic patterns of the cases. According the last column of Table 4 shows these patterns. Each pattern has a title according to the most shared actions between the close patterns cases. The patterns adopted by the airlines were hybrid 1: do all, which was adopted by one case. Hybrid 2: design-standards-operating, which was adopted by two cases. Conformance with ICAO Standards, which was adopted by four cases. Hybrid 3: design-operating, which was adopted by one case. Hybrid 4: design-airport standards, which was adopted by one case. Flights operating, which was adopted by two case. Hybrid 5: design- ICAO standards, which was adopted by two cases. Conformance with FAAS Standards, which was adopted by one case. Aircraft design, which was adopted by two cases.

Hybrid1: Do all Strategic Pattern

Table 6 shows hybrid1: do all strategic pattern. It can be seen that this pattern was adopted by one case, which was E2. According to this pattern, the taken actions categories were aircraft design, aircraft conformance with ICAO noise standards, conformance with airport fly quiet program, and flights operations. The fleets were renewed. In addition, 97.7% of thc aircrafts laid out ICAO's Ch.4. This airline was the 1st ranked according to the noise performance (long haul) of Heathrow fly quiet. This airline adopted the night flight and the day flight rescheduling and procedure improvement according to CAD procedure, in addition performed well with respect to the continuous descent approach (CDA).

Hybrid 2: Design-Standards-Operating Strategic Patterns

Table 7 shows hybrid 2: design-standard-operating strategic patterns. It can be seen that this pattern was adopted by two cases (E1&E3). There is some

Table 5. Noise Management Actions Indices and Strategy Patterns

Case ID	Aircraft design (D)	Aircraft conformance with ICAO Noise Standards (ICAOS)	Aircraft conformance with FAA noise standards (FAAS)	conformance with Airport fly quiet program (AS)	Flights operating (FO)	Strategy Pattern	Strategy pattern title
E2	2.50 (M)	5.00 (H)	0.00	5.00 (H)	3.75 (H)	D-ICAOS-AS-FO	Hybrid 1: Do all
E1	5.00 (H)	5.00 (H)	0.00	0.00	5.00 (H)	D-ICAOS-FO	Hybrid 2: Design-Standards-Operating
E3	2.5 (M)	5.00 (H)	0.00	0.00	2.8 (M)	D-ICAOS-FO	Hybrid 2: Design-Standards-Operating
ME2	0.00	5.00 (H)	0.00	0.00	1.25 (L)	ICAOS	Conformance with ICAO Standards
A2	2.50 (M)	0.00	0.00	0.00	2.50 (M)	D-FO	Hybrid 3: Design-Operating
E6	2.50 (M)	0.00	0.00	5.00 (H)	1.25 (L)	D-AS	Hybrid 4: Design- Airport Standards
A6		0.00	0.00	0.00	5.00 (H)	FO	Flights operating
NA5	2.50 (M)	5.00 (H)	0.00	0.00	0.00	D-ICAOS	Hybrid 5: Design- ICAO standards
SA1	2.50 (M)	5.00 (H)	0.00	0.00	0.00	D-ICAOS	Hybrid 5: Design- ICAO Standards
NA3	0.00		5.00 (H)	0.00	1.25 (L)	FAAS	Conformance with FAA Standards
NA2	0.00	5.00 (H)	0.00	0.00	0.00	ICAOS	Conformance with ICAO Standards
A3	0.00	5.00 (H)	0.00	0.00	0.00	ICAOS	Conformance with ICAO Standards
A7	0.00	5.00 (H)	0.00	0.00	0.00	ICAOS	Conformance with ICAO Standards
NA6	2.50 (M)	0.00	0.00	0.00	0.00	D	Aircraft Design
ME1	0.00	0.00	0.00	0.00	2.50 (M)	FO	Flights Operating
A5	2.50 (M)	0.00	0.00	0.00	0.00	D	Aircraft Design
ME3	2.50 (M)	0.00	0.00	0.00	0.00	D	Aircraft Design
E4	0.00	0.00	0.00	0.00	1.25 (L)	--------	----------

(L): Low **(M):** Moderate **(H):** High

Table 6. Hybrid1: Do all Strategic Patterns

Case ID	Strategy Pattern	Actions category Action Index	Actions taken
E2	D-ICAOS-AS-FO	**Aircraft Design (D)** Action Index= 2.5 (M)	Fleet renewal (B737, B737 MAX, A320, A350, B787)
		Aircraft conformance with ICAO Noise Standards (ICAOS) Action index= 5.00 (H)	97.7% of aircraft laid out the ICAO's Ch.4
		Conformance with Airport fly quiet program (AS) Action Index= 5.00 (H)	1st ranked according to noise performance (short haul fleet) of Heathrow fly quiet program
			6th and 10th ranked according to noise performance (long haul fleet) of Heathrow fly quiet program
		Flights operating (FO) Action Index=3.75 (H)	Night flight rescheduling and procedures improvement according to CAD procedures
			Day flight procedures improvement according to CAD procedures
			Performed well with respect to continuous descent approach (CDA)

(L): Low **(M):** Moderate **(H):** High

shared actions between the two cases related to the aircraft design and the aircrafts conformance with ICAO noise standards. Both cases adopted the fleet renewal, also, 100% of E1 aircrafts and 90% of E3 aircrafts were conformed with ICAO noise standards. However, each case adopted different flights' operating actions. The first case adopted the night flight rescheduling and the procedures improvement according to CAD procedures, improve the departure and the approach procedures with civil aviation authority abatement procedure 2, and the area navigation (RNAV) departure. However, the second case performed well with respect to the continuous descent approach (CDA),

Table 7. Hybrid 2: Design-Standard-Operating Strategic Patterns

Case ID	Strategy pattern	Action category Action Index	Actions taken
E1	D-ICAOS-FO	**Aircraft Design (D)** Action Index= 5.0 (H)	Fleet renewal (737, 737 MAX, A320, A350, B787)
			Modification of aircraft design to lower noise ambient as winglets design
		Aircraft conformance with ICAO Noise Standards (ICAOS) Action Index= 5.0 (H)	100% of aircraft laid out the ICAO's Ch.4
		Flights operating (FO) Action Index=5.0 (H)	Night flight rescheduling and procedures improvement according to CAD procedures
			improve departure and approach procedures with civil aviation authority abatement procedure 2
			electrical taxiing
			Area navigation (RNAV) departure
E3	D-ICAOS-FO	**Aircraft Design (D)** Action Index= 2.5 (M)	Fleet renewal (737, 737 MAX, A320, A350, B787)
		Aircraft conformance with ICAO Noise Standards (ICAOS) Action Index= 5.0 (M)	90% of aircraft laid out the ICAO's Ch.4
		Flights operating (FO) Action Index=2.80 (M)	Performed well with respect to continuous descent approach (CDA)
			Rescheduling takeoffs and landing at less undesirable time from a noise perspective

(L): Low **(M):** Moderate **(H):** High

and rescheduled the takeoffs and the landing at less undesirable time from a noise perspective.

Conformance with ICAO Standards Strategic Patterns

Table 8 shows the conformance with ICAO standards strategic patterns. It can be seen that this pattern was one of the widely adopted patterns by the airlines. It was adopted by four airlines (ME2, NA2, A3, &A7). The shared actions was the conformance with ICAO's Ch.4. 100% of the aircrafts of ME2, 94% of NA2 aircrafts and 100% of A7 aircrafts satisfied ICAOs Ch.4. However, other actions have been adopted by ME2, NA2 and A3. 8% of ME2 aircrafts were met ICAO's Ch.14. 100% of NA2 aircrafts satisfied ICAO's Ch.3 and 100% of A3 aircrafts satisfied ICAO's Ch.4.

Table 8. Conformance with ICAO Standards Strategic Patterns

Case ID	Strategy pattern	Action category Action Index	Actions taken
ME2	ICAOS	**Aircraft conformance with ICAO Noise Standards (ICAOS)** Action Index= 5.0 (H)	100% of aircraft laid out the ICAO's Ch.4
			8% of aircraft meet ICAO's Ch.14
NA2	ICAOS	**Aircraft conformance with ICAO Noise Standards (ICAOS)** Action Index= 5.0 (H)	94% of aircraft laid out the ICAO's Ch.4
			100% of aircraft laid out the ICAO's Ch.3
A3	ICAOS	**Aircraft conformance with ICAO Noise Standards (ICAOS)** Action Index= 5.0 (H)	100% of aircraft laid out the ICAO's Ch.4.
A7	ICAOS	**Aircraft conformance with ICAO Noise Standards (ICAOS)** Action Index= 5.0 (H)	100% of aircraft laid out the ICAO's Ch.4

(L): Low **(M):** Moderate **(H):** High

Hybrid 3: Design-Operating Strategic Pattern

Table 9 shows hybrid 3: design-operating strategy pattern. It can be seen that this pattern was adopted by A2. The taken actions by this case were the fleet renewal, the controlling and the monitoring of noise from the airports and the optimum flights route.

Table 9. Hybrid 3: Design-Operating Strategic Pattern

Case ID	Strategy pattern	Action category Action Index	Actions taken
A2	D-FO	**Aircraft Design (D)** Action Index= 2.5 (M)	Fleet renewal (737, 737 MAX, A320, A350, B787)
		Flights operating (FO) Action Index=2.5 (M)	Controlling and monitoring noise from airports
			Optimum flights routes

(L): Low **(M)**: Moderate **(H)**: High

Hybrid 4: Design- Airport Standard Strategic Pattern

Table 10 shows hybrid 4: design-airports standards strategy pattern. It can be seen that this pattern was adopted by E6. The taken actions by this case were the fleet renewal, and the satisfying of Heathrow (short haul) fly quiet standards. This airline was the 1st ranked according to Heathrow airport fly quiet program.

Table 10. Hybrid 4: Design- Airport Standards Strategic Pattern

Case ID	Strategy pattern	Action category Action Index	Actions taken
E6	D-AS	**Aircraft Design (D)** Action Index= 2.5 (M)	Fleet renewal (B737, B737 MAX, A320, A350, B787)
		Conformance with Airport fly quiet program (AS) Action Index= 5.00 (H)	1st ranked according to noise performance (short haul fleet) of Heathrow fly quiet program

(L): Low **(M)**: Moderate **(H)**: High

Flight Operating Strategic Pattern

Table 11 shows flight operations strategy pattern. It can be seen that this pattern was adopted by two cases A6 and ME1. Each case adopted a different combination of the flight operating actions. A6 adopted the continuous descent approach, the controlling and the monitoring noise from the airport, practicing the noise abatement procedures during takeoff, and practicing the noise abatement procedures during the landing. However, ME1 adopted the optimum flights routes and reduced using of APU.

Table 11. Flight Operating Strategic Pattern

Case ID	Strategy pattern	Action category Action Index	Actions taken
A6	FO	**Flights Operating (FO)** Action Index=5.0 (H)	Performed well with respect to continuous descent approach (CDA)
			Controlling and monitoring noise from airports
			Practicing noise abatement procedures during takeoff
			Practicing noise abatement procedures during landing (reduce flap, delayed flap and idle reserve)
ME1	FO	**Flights Operating (FO)** Action Index=2.5 (M)	Optimum flights routes
			Reduce using of APU

(L): Low **(M):** Moderate **(H):** High

Hybrid 5: Design-ICAO Standards Strategic Pattern

Table 12 shows hybrid 5: design-ICAO standards strategy pattern. It can be seen that this pattern was adopted by two cases NA5 and SA1. The shared actions between these two patterns were satisfying ICAO's Ch.4. 8% of NA5 aircrafts, and 100% of SA1 aircrafts satisfied ICAO's Ch.4. However, each case adopted different actions. NA5 adopted modifying the aircraft design to a lower noise ambient as winglets design, also 100% of its aircrafts laid out the ICAO's Ch.3. On the other hand, SA1 adopted a different design action, which was the fleet renewal.

Table 12. Hybrid 5: Design- ICAO Standards Strategic Pattern

Case ID	Strategy pattern	Action category Action Index	Actions taken
NA5	D-ICAOS	**Aircraft Design (D)** Action Index= 2.5 (M	Modification of aircraft design to lower noise ambient as winglets design
		Aircraft conformance with ICAO Noise Standards (ICAOS) Action Index= 5.0 (H)	8% of aircraft laid out ICAO's Ch.4
			100% of aircraft laid out the ICAO's Ch.3
SA1	D-ICAOS	**Aircraft Design (D)** Action Index= 2.5 (M	Fleet renewal (B737, B737 MAX, A320, A350, B787)
		Aircraft conformance with ICAO Noise Standards (ICAOS) Action Index= 5.0 (H)	100% of aircraft laid out the ICAO's Ch.4

(L): Low **(M)**: Moderate **(H)**: High

Conformance With FAA Standards Strategic Pattern

Table 13 shows conformance with FAA standards strategy pattern. It can be seen that this pattern was adopted by NA3. The taken action by this case was satisfying FAA stage 4 criteria, 78% of NA3 aircrafts satisfied this standard.

Table 13. Conformance with FAA Standards Strategic Pattern

Case ID	Strategy pattern	Action category Action Index	Actions taken
NA3	FAAS	**Aircraft conformance with FAA Noise Standards (FAAS)** Action Index= 5.0 (H)	78% of aircrafts meet stage 4 criteria of FAA

(L): Low **(M)**: Moderate **(H)**: High

Aircraft Deign Strategic Pattern

Table 14 shows aircraft design strategy pattern. It can be seen that this pattern was adopted by NA6, A5 and ME3. The taken action by these case was the fleet renewal.

Table 14. Aircraft Design Strategic Pattern

Case ID	Strategy pattern	Action category Action Index	Actions taken
NA6	D	**Aircraft Design (D)** Action Index= 2.5 (M)	Fleet renewal (B737, B737 MAX, A320, A350, B787)
A5	D	**Aircraft Design (D)** Action Index= 2.5 (M)	Fleet renewal (B737, B737 MAX, A320, A350, B787)
ME3	D	**Aircraft Design (D)** Action Index= 2.5 (M)	Fleet renewal (B737, B737 MAX, A320, A350, B787)

(L): Low **(M)**: Moderate **(H)**: High

DISCUSSION

The widely adopted strategic pattern was hybrid, which was adopted by seven cases. This indicates that the airlines are trying to adopt the principles of balanced approach of managing noise. In order to have the outcomes for noise reduction, it is recommended to adopt more action categories. On the other hand, the widely shared action categories between the hybrid strategic patterns was design (the most action taken by the airlines was the fleet renewal). The other shared actions category was the conformance with ICAO's standards (Ch.3&4) and the operating actions. The design action is the main factor in reducing the noise, this is called by ICAO's balanced approach the reduction of noise at source (ICAO, 2019b). Making this decision by the airlines is logical since the aviation noise is caused by air passing over the aircraft body, which causes the turbulence and the friction, or the noise is caused by the sounds of engine moving parts and the sound of air being expelled at the high speed (Glasgow Airport, 2014).

Regarding the wide adopting of ICAO standards (Ch.3 &4) this is justified since these standards have been mandatory for the airlines. Also, the adopting of these standards is a logical outcome of renewing the fleets, since the new aircrafts are much quieter than that manufactured before 20, 30 or 40 years ago. For example, the noise footprint of B787 Dreamliner is 60% smaller than similarly sized aircrafts (Glasgow Airport, 2014), also the last version of B747, B747-8 intercontinental claim a 30% noise performance improvement over that of its predecessor B747-400 and A380 is one quietest wide-body jet aircraft (Civil Aviation Authority, 2014). A380 powered by Rolls-Royce Trent 900 and B787 powered by Rolls-Royce Trent 1000 are more quiet than the predecessors are. The engines of these aircrafts increased engine bypass, rations, and nacelles with zero-splice intake liners (Sustainable Aviation, 2018).

The widely adopted operational actions by the airlines were numerous as the night flight rescheduling and the procedures improvement according to CAD procedures, the day flight procedures improvement according to CAD procedures, the improving of departure and approach procedures with the civil aviation authority abatement procedure 2, the electrical taxiing, the controlling and the monitoring noise from the airports, the optimum flights routes..etc. This is consistent with the third dimension of the ICAO balanced approach, which is the noise abatement operational procedures. According to this dimension, the daily operating of aircrafts as the preferential runway, the routes, the noise abatement procedures fora take-off and the landing will affect the noise (ICAO, 2019d). In addition, consistent with ICAO's fourth dimension, the operating restrictions, according to this dimension these restriction are related to the operating of airports, and include the curfews, the night time restrictions, the noise quotes/budgets, the cap rules (ICAO, 2019e). According to the previous studies, the noise is affected by the flight events (departure and arrival) and the reconstruct of aircraft flight path (Pretto, Giannattas, De Gennavo & Zanon, 2019). The noise reduction of -1 EPNL is possible with a 450m landing and within a 3.5-degree glide slope (Fillppone, 2017). In addition, scheduling the landing of a sequence of the aircrafts will minimize the noise (Rodrignes-Diaz, Adenso-Diaz & Gonzalez-Torre, 2019). Moreover, the actual take-off weight and both the aircraft ground speed and altitude are the most significant factors that affected the noise emissions level (Gagliardi, Teti & Licitra, 2018).

CONCLUSION

The aim of this chapter was to report the effective airlines noise management strategies. This research issue has been raised by limited studies, most of previous studies have investigated the abatement operations procedure, the reduction of noise at source and other studies have investigated the land-use planning. Very limited studies have investigated the airlines green operations strategies, these studies have investigated the general strategic patterns. The strategic patterns of airlines noise management have not been investigated by the previous studies. The main results of this study were the noise management strategic patterns adopted by airlines were seven hybrid patterns, four conformance with ICAO standards patterns, two flights operating patterns, one conformance with FAA Standards pattern, and three aircraft design patterns.

The widely shared action categories between the hybrid strategic patterns was design (the most action taken by the airlines was the fleet renewal). The other shared actions category was the adopting of ICAO's standards (Ch.3&4) and the operating actions. The widely adopted operational actions by the airlines were numerous as the night flight rescheduling and the procedures improvement according to the CAD procedures, the day flight procedures improvement according to the CAD procedures, improving the departure and the approach procedures with civil aviation authority abatement procedure 2, the electrical taxiing, the controlling and the monitoring of noise from the airports, and the optimum flights routes.

The results of this chapter will benefit the academics and the industry experts alike. The academics have in-depth reporting of the green operations actions to reduce the noise. Therefore, they can develop hypotheses for the future researchers. The experts in the industry have a clear idea about the different strategy patterns that could be adopted. Despite the significant contribution of this chapter, some recommendations could be given for the future researchers. It is recommended to investigate a larger sample size to have more valid results, it is recommended to investigate the interrelationship between the actions categories and the actions taken to have a better understanding for the independent, the moderating, the mediating and the dependent variables. It is recommended to examine the effectiveness of the strategic patterns by investigating the impact of green actions on the noise reduction.

REFERENCES

Airport Council International (2010). *ACI aircraft noise rating index.* Author.

Butcher, L. (2017). *Aviation noise.* House of Community Library, London. https://researchbriefings.parliament.uk/ResearchBriefing/Summary/SN00261#fullreport

CANSO. (2015). *Managing the impact of aviation: a guide for airport operations and air navigation service providers.* https://www.canso.org/sites/default/files/Managing%20the%20Impacts%20of%20Aviation%20Noise_HQ.pdf

Civil Aviation Authority. (2014). *Managing aviation noise*. https://publicapps.caa.co.uk/docs/33/CAP%201165%20Managing%20Aviation%20Noise%202.pdf

Filippone, A. (2017). Options for aircraft noise reduction on arrival and landing. *Aerospace Science and Technology*, *60*, 31–38. doi:10.1016/j.ast.2016.10.027

Flores, R., Asensio, C., Gagliardi, P., & Licitra, G. (2019). Study of the correction factors for aircraft noise façade measurements. *Applied Acoustics*, *145*, 399–407. doi:10.1016/j.apacoust.2018.10.007

Gagliardi, P., Teti, L., & Licitra, G. (2018). A statistical evaluation on flight operational characteristics affecting aircraft noise during take-off. *Applied Acoustics*, *134*, 8–15. doi:10.1016/j.apacoust.2017.12.024

Glasgow Airport. (2014). *Noise action plan 2013-2018*. https://www.glasgowairport.com/media/2067/current-noise-action-plan.pdf

GRI. (2017). *About GRI*. https:// www.globalreporting.org/Information/about-gri/Pages/default.aspx

IATA. (2013). *IATA position on noise-related operating restrictions*. https://www.iata.org/policy/environment/Documents/paper-on-operating-restrictions-august-2013.pdf

IATA. (2018). *Night flights facts sheet*. https://www.iata.org/pressroom/facts_figures/fact_sheets/Documents/fact-sheet-night-flights.pdf

ICAO. (2019a). *Aircraft noise*. https://www.icao.int/environmental-protection/Pages/noise.aspx

ICAO. (2019b). *Reduction noise at source*. https://www.icao.int/environmental-protection/pages/Reduction-of-Noise-at-Source.aspx

ICAO. (2019c). *Land-use planning and management*. https://www.icao.int/environmental-protection/pages/Land-use-Planning-and-Management-.aspx

ICAO. (2019d). *Noise-abatement operational procedures*. https://www.icao.int/environmental-protection/pages/Noise-Abatement-Operational-Procedures.aspx

ICAO. (2019e). *Operating restrictions*. https://www.icao.int/environmental-protection/Pages/Operating-Restrictions.aspx

London Assembly Environment Committee. (2019). *Aircraft noise*. https://www.london.gov.uk/sites/default/files/aircraft_noise_report.pdf

Migdadi, Y. K. A. A. (2018). Identifying the best practices of airlines' green operations strategy: A cross-regional worldwide survey. *Environmental Quality Management*, *28*(1), 21–32. doi:10.1002/tqem.21575

Migdadi, Y. K. A. A. (2019). Identifying the effective taxonomies of airline green operations strategy. *Management of Environmental Quality*.

Migdadi, Y. K. A. A. (2020). Green Operational Strategy for Airlines: Content and Regional Analysis. In Cases on Green Energy and Sustainable Development (pp. 193-229). IGI Global.

Pretto, M., Giannattasio, P., De Gennaro, M., Zanon, A., & Kühnelt, H. (2019). Web data for computing real-world noise from civil aviation. *Transportation Research Part D, Transport and Environment*, *69*, 224–249. doi:10.1016/j.trd.2019.01.022

Rodríguez-Díaz, A., Adenso-Díaz, B., & González-Torre, P. L. (2019). Improving aircraft approach operations taking into account noise and fuel consumption. *Journal of Air Transport Management*, *77*, 46–56. doi:10.1016/j.jairtraman.2019.03.004

Sustainable Aviation. (2018). *The SA noise road-map: a blueprint for managing noise from aviation sources to 2050*. https://www.sustainableaviation.co.uk/wp-content/uploads/2018/06/SA-Noise-Road-Map-Report.pdf

Torija, A. J., Roberts, S., Woodward, R., Flindell, I. H., McKenzie, A. R., & Self, R. H. (2019). On the assessment of subjective response to tonal content of contemporary aircraft noise. *Applied Acoustics*, *146*, 190–203. doi:10.1016/j.apacoust.2018.11.015

Chapter 7
Airline Effective Green Operations Strategy Patterns:
Case Level Analysis

ABSTRACT

This chapter reports the effective green operations strategy patterns. To achieve this aim, the green operations strategy patterns of 23 airlines from five regions were investigated. This chapter depended on the results of previous chapters. Accordingly, the reported effective green operations strategy under each indicator was used to report the effective green operations strategy patterns. The special tailored data analysis techniques were used for this purpose. The chapter reported eight effective strategy patterns, the strategy patterns were water management, waste management+, GHG and noise management, GHG emissions, GHG waste and noise, GHG and water management+, do all and noise management. This chapter helps the decision makers and the academics alike since all effective patterns were identified.

INTRODUCTION

The patterns of green operations strategy are the strategy configurations that adopted by airlines. These patterns could be developed by using the typologies or the taxonomies approach (Migdadi, 2011a, 2013). Reporting the effective green operations strategy patterns is very important since it combines the effective actions, and green indicators in one construct. Accordingly, the most

DOI: 10.4018/978-1-7998-4255-2.ch007

effective patterns could be adopted. Developing green operations strategy by using the taxonomies approach is more realistic, since it is developed by using quantitative data to construct the different strategy configuration. However, thc typology approach seeks to develop more ideal and base theory of the effective green operations strategy.

Reporting the effective taxonomies of airlines operations strategy in general and green operations strategy in particular is still limited. For example, some of the previous studies have reported the effective capacity strategy (e.g. Alkhatib & Migdadi, 2018) and the effective scheduling strategy of airlines (e.g. Alkhatib & Migdadi, 2019). The effective green operations strategy contribution have been reported by some studies as the study of Migdadi (2018, 2019). The limited studies have adopted the clustering analysis technique by using K-means clustering analysis. The previous studies have reported the strategy patterns, with limited in-depth analysis for the effective green operations strategies adopted under each green indicators.

This chapter bridged the gaps of previous studies by reporting the effective green operations strategy patterns adopted by airlines. This chapter adopted the case level analysis, and depended on the results of previous chapters. Accordingly, the main question of this chapter was: what were the most effective green operations strategy patterns that adopted by airlines? To answer the question the following objectives were realized:

1. Identifying the effective green operations strategy adopted by each case under each green indicator.
2. Rating the green operations strategy adopted by each airline under each green indicator.
3. Identifying the green operations strategy patterns that adopted by airlines.
4. Rating the performance of each green operations strategy patterns that adopted by airlines.

This chapter is arranged in sections: the first one is the literature review, followed by the methodology and methods, then, the results of the empirical study, next, the discussion, and the final section is the conclusions.

LITERATURE REVIEW

This part of the chapter discussed the key literature of the green operations strategy patterns. The first part of the literature aimed to discuss how the operations strategy configurations were developed. Accordingly, the different approaches that adopted by the previous studies were discussed. The second part of the literature review discussed the green operations configurations, with emphasis on the literature related to airlines industry.

Operations Strategy Configuration Approaches

Strategy configuration is a building block of attributes and actions, these configurations are called strategy patterns. Two approaches have been adopted to develop operations strategy configurations: the first one is the typology approach, and the second is the taxonomies approach. According to the typology approach, the strategy configurations are ideal models or patterns that could be developed by investigating a limited number of best practices, or it could be developed by using the literature. The typologies are used as a reference for comparing the practices in order to modify the practices based on the typology attributes. In addition, the typologies could be used as a classification scheme for the practices, and the typologies could be used as initial or a ground theory which will be used later for more extended quantitative research (Migdadi, 2011a, 2013).

The second strategy configuration approach is the taxonomies of strategy. It is a quantitative approach, based on quantitative analysis of the collected facts about the investigated cases. Accordingly, the taxonomies approach results present both the effective and the ineffective strategy patterns. Therefore, it is more realistic in comparison with the typologies approach (Migdadi, 2011a). The widely adopted quantitative analysis approach by the previous studies was the clustering analysis in general, and K-means clustering analysis in particular (e.g. Migdadi, 2011b, Migdadi, 2012a; Migdadi, 2014a; Alkhatib & Migdadi, 2018; Migdadi & Elzaqaiba, 2018; Migdadi, 2018; Alkhatib & Migdadi, 2019; Migdadi, 2019; Migdadi & Omari, 2019).

However, other analysis techniques rather than K-means were adopted to classify the cases into clusters. For example, Migdadi (2014) ranked cases according to the realized change in volume flexibility by computing the differences between each case and preceding ranked case, then, finding the average difference, which will be used as a cutting point between clusters, if the difference in volume flexibility is more than the average difference in volume flexibility, the new cluster will be recognized. In addition, the study of Migdadi (2012b) used the results of customers survey in order to identify the order winners criteria of mobile phone service providers, then, the significant operational actions taken by leaders in comparison with followers and laggards were identified, finally the indicative models were developed by using a special tailored quantitative approach, since the sample size was three cases.

Usually multi-attributes have been used to generate clusters or strategy groups (e.g. Migdadi, 2012a; Migdadi, 2014a; Alkhatib & Migdadi, 2018; Migdadi & Elzaqaiba, 2018; Migdadi, 2018; Alkhatib & Migdadi, 2019; Migdadi, 2019; Migdadi & Omari, 2019). However, in some cases single attribute has been used (e.g. Migdadi, 2014b). Using multi-attributes are more preferred than single, since the adopting of quantitative approach as K-means will be easier, but this requires the adopting of all attributes by all cases, in other word, collecting quantitative facts for all attributes and for each case. In some situations, the attributes have been not adopted by the case at all, the data is missing (e.g. Migdadi, 2018), or the study focusing on one attribute (e.g. Migdadi, 2014), in this case the researchers have two choices, whether to assume the existence of an artificial variable, and conduct clustering analysis for two variables, the actual and the artificial (e.g. Migadadi, 2018), or develop an innovative and a special tailored data analysis technique (e.g. Migdadi, 2014b).

The attributes adopted by previous studies were the operations competitive priorities or capabilities, the performance indicators, and the operational strategic actions. For example, Migdadi (2017) classified mobile phone service providers into four strategic groups based on the operational competitive priorities as the cost of calls service, the diversity of entertainment services, the diversity of offers, and the service accessibility. Alkhatib & Migdadi (2019), developed the taxonomies of airline scheduling strategies based on the operational actions as the number of destinations, the number of continents, the number of hubs, the ranges of flights and the international orientation. In addition, Alkhatib & Migdadi (2018) developed the taxonomies of airline capacity strategies based on the operational actions as the number of destinations, the number of countries, the aircrafts diversity, the fleet size, the average number of employees per aircraft, the total number of employees, the average number of seats per aircraft, and the total number of seats. Migdadi (2014) developed the taxonomies of effective volume flexibility of hospitals by using the volume flexibility as an operational competitive capability, and identified the significant actions later for the most effective cluster of hospitals.

Migdadi (2016) developed the taxonomies of green operations strategy of mobile phone service providers based on green indicators as the percentage of CO2 emissions, the percentage of solid wastes, the percentage of hazard wastes, the percentage of recycling, the percentage of water consumption, and the percentage of energy consumption. Migdadi & Elzagaiba (2018) developed the taxonomies of green manufacturing strategies based on the financial indicators, the operational indicators, and the environmental indicators. Migdadi & Omari (2019) developed the taxonomies of green operations strategy of hospitals based on the green indicators as the percentage of paper consumption, the fuel consumption, the hazard wastes, the electrical power consumption, the non-hazard wastes, the direct greenhouse gases emissions and the indirect gases emissions. Migdadi (2018, 2019) developed the taxonomies of airlines green operations strategy by using green indicators as the direct GHG1 emissions, the indirect GHG2 emissions, the energy consumption, the fuel consumption, the waste generated, the water consumption, and the recycling of wastes.

Green Operations Strategy Configurations

The research contribution of the green operations strategy configuration in general, and the green operations configuration of airlines in particular is still limited. The previous studies have focused on developing the taxonomies by adopting K-means clustering analysis, and by using the green operational capabilities as attributes for developing the configurations (e.g Migdadi, 2016; Migdadi & Elzaqaiba, 2018; Migdadi & Omari, 2019; Migdadi, 2019a), rather than by using the green operational actions. The number of strategy patterns reported by the previous studies varies, some of them have developed four taxonomies of the green operations strategy (e.g. Migdadi & Omari, 2019; Migdadi, 2019), and other studies have developed three taxonomies of the green operations strategy (e.g. Migdadi, 2016; Migdadi & Elzaqaiba, 2018).

The green taxonomies of hospitals operations strategy were the resources & waste management (a high reduction in paper use, a moderate reduction in fuel consumption and hazard wastes), the electrical power management (a moderate reduction in electricity consumption), the non-hazard waste (a high reduction in non-hazard wastes), and the emissions & resources management (high in reducing paper use, high in reducing fuel consumption and high in reducing direct greenhouse emissions) (Migdadi & Omari, 2019). However, the taxonomies of green mobile phone producers operations strategy were the hazard material management (a high reduction in hazard wastes), the solid waste & recycling (a high reduction of solid waste and a high increase in recycling), and hybrid (hazard wastes and solid & recycling) (Migdadi,

2016). The green strategy patterns of manufacturing sector were the caretaker (achieved a low to moderate level of the financial, the operational and the environmental indicators), agile (high achievement for all indicators), and lean (achieved a moderate level of all indicators) (Migdadi & Elzaqaiba, 2018).

The green operations strategy patterns of the airlines were the low to moderate emphasizer (a moderate reduction of GHG2 emissions, a low reduction for fuel consumption, a low reduction for energy consumption, and a moderate reduction of wastes), the misleading strategy (no improvement in green performance), the high emphasizer (a high reduction in GHG1 emission, and fuel consumption), and the low emphasizer (extremely a low reduction in GHG1, GHG2 emissions, and a low reduction of energy consumption) (Migdadi, 2019). Moreover, the airlines green strategy patterns were strategy group 1 (reducing GHG emissions, reducing wastes and reducing resources consumption), strategy group 2 (reducing GHG emissions, and reducing resources consumption), strategy group 3 (reducing wastes and reduce using of resources), strategy group 4 (reducing wastes) (Migdadi, 2018).

There are shared strategic green patterns between the different sectors. For example, the low emphasizer of airlines green strategy that reported by Migdadi (2019), is close to the caretaker manufacturing strategy pattern that reported by Migdadi & Elzagaiba (2018). The low to moderate emphasizer strategic group of airlines is close to the resources & waste management strategy group of hospitals that reported by Migdadi & Omari (2019). The waste reduction strategic group of airlines that reported by Migdadi (2018), is close to the solid waste and recycling strategic group of mobile phone producers that reported by Migdadi (2016). Reducing GHG emissions and reducing resources consumption strategic group of airlines that reported by Migdadi (2018), is close to the emissions & resources management strategic group of hospitals that reported by Migdadi & Omari (2019).

Different significant green operations actions were taken by the different strategy patterns. For example, the low emphasizer strategy group of airlines took three actions categories as adopting the energy saving actions, the using of sustainable energy, and the upgrading of facilities. However, the low to moderate emphasizer strategy group adopted fuel management action to reduce the fuel consumption. Moreover, the change in engine design, improve the operating of engines and adopting LEED to save the energy and reduce GHG2 emissions. Further, three actions took to reduce the wastes as recycling, upcycling and reusing of on-board and the ground wastes. On the other hand, the actions took by the airlines of high emphasizer strategy group were two actions to save the fuel and reduce GHG1 emissions, these actions were the aircraft design, and the fuel management (Migdadi, 2019).

METHODOLOGY AND METHODS

This part of the chapter discussed how the strategic patterns were identified, and how the effectiveness of these patterns were assessed. In this regards, the design of sample and the sources of data were discussed. Then, how the data was analyzed to formulate strategy patterns, and assess the effectiveness of these patterns. Finally, supportive examples were presented in the appendices.

Sample Design

The unit of analysis for this chapter is the case (airline); accordingly, 23 airlines which adopted the effective green operations strategy were the sample of the study. The sample is a purposive accordingly to adopting of the effective green operations strategy. The distribution of sample across regions is as presented in Table 1. It can be seen that seven cases were surveyed in Asian, seven cases were surveyed in Europe, however, five cases were surveyed in North America, three cases in the Middle East and one case in South America.

Table 1. Sample of study

Regions	Samples size
A: Asia	7
E: Europe	7
NA: North America	5
ME: Middle East	3
SA: South America	1
Total	23

Sources of Data

This chapter depended on the analysis outcomes of the previous five chapters. Accordingly, the data required for this chapter was the effective strategy patterns adopted by each case to achieve each green performance indicator. At the end of each chapter there is a table summarizes the effective green strategy patterns adopted by the airlines, this table was the source of data that used in this chapter.

Data Analysis

Rate the Performance of Green Strategies Adopted by Each Case

The following steps were taken to rate the performance:

Step1: identifying the case achieved the best performance.
Step2: Dividing the performance that was achieved by each case by the best. The result is the relative performance
Step3: Multiplying the relative performance by 5 (the maximum rating value of the scale). The result is the rating value of performance.
Step 4: Assign a level of performance for each case according to the achieved rating value by using the following classification scheme. The assigned level represents the rank of case performance.

From more than 0 to 1.67 is low level. From more than1.67 to 3.33 is moderate level. From more than 3.33to 5.00 is high level.

Step 5: Computing the average rated performance of the green strategies that were adopted by each case

Please see the example in the Appendix, Tables 4 and 5.

Identify the Green Strategy Patterns

The configuration of strategy patterns could be identified based on the actions or the green operational capabilities. In this chapter, the strategy patterns were identified by using the green strategy patterns that were adopted by each case. The base of developing the green strategy patterns of each case was the significant actions that were adopted by each case. On the other hand, the widely adopted analysis approach by the previous studies to build the strategy configurations was the clustering analysis techniques as K-means clustering analysis. However, the adopting of such technique requires the adoption of all attributes by all cases. In the case of this study, not all effective patterns were adopted by all cases, so it is impossible to use the K-means clustering analysis. In addition, the effectiveness of some strategy patterns as the noise management patterns have not been assessed, since the secondary published data is unavailable for this indicator.

Accordingly, the following specific tailored technique was adopted:

Step 1: identifying the green operations strategy pattern adopted by each case.
Step 2: Identifying the shared green operations strategy patterns across cases and assign a title reflects the maximum-shared patterns.
Step 3: Computing the average rated performance of the pattern.

Please see the example in the Appendix 2, Tables 6 and 7.

RESEARCH RESULTS

This part of this chapter presented the main results of the empirical study. The aim of this part was to identify the effective green operations strategy patterns. In order to facilitate the reading of the strategy patterns alternatives, the following table (Table 2 shows the acronyms of the strategy patterns alternatives).

Effective Green Operations Strategy Patterns

Table 3 shows the effective green operations strategy patterns that were adopted by the airlines. It can be seen that the number of strategy patterns adopted by the airlines was eight. These patterns were water management, waste management+, GHG & noise management, GHG emissions, GHG waste & noise, GHG & water management+, do all, and noise management. The performance of these patterns was moderate for waste management+, GHG & noise management, GHG emissions, GHG waste & noise, and GHG & water management+, and low for water management, and do all. The most adopted pattern by the airlines was GHG & noise management, which was adopted by five cases. Four cases adopted the noise Management pattern. The lowest adopted pattern was water management and do all pattern. However, the remaining patterns were adopted by two airlines.

Table 2. List of Acronyms Used in Presenting the Chapter Results

Acronym	Interpretation
A	Facilities and building accreditation
AS	Conformance with Airport fly quiet program
D	Aircraft Design
ES	Facilities and building energy saving
F	Flights fuel management
FAAS	Aircraft conformance with FAA noise standards
FO	Flights operating
FWR	Facilities and buildings water recycling and recovery
FWS	Facilities and buildings water saving
GHG1	Direct greenhouses gas emission
GHG2	Indirect greenhouse gas emission
HRU	Hazard wastes reusing
HUP	Hazard wastes upcycling
HRC	Hazard wastes recycling
ICAOS	Aircraft conformance with ICAO Noise Standards
IRC	Industrial wastes recycling
IRU	Industrial wastes reusing
IUP	Industrial wastes upcycling
M	Flights maintenance management
MWR	Maintenance water recycling and recovery
MWS	Maintenance water saving
NM	Noise management
O	Flight Operating
ORC	On-board and ground wastes recycling
ORP	On-board and ground reduce using of paper
OUP	On-board and ground wastes upcycling
R	Flights routes management
RF	Upgrading and replacing facilities
SE	Facilities and building sustainable energy
VD	Engine and vehicle design
VO	Engine and vehicle Operating
VM	Engine and vehicle maintenance
W	Flights weight management
WSM	Waste management
WTM	Water Management

Table 3. Airline Effective Green Operations Strategy Patterns

Pattern	Case ID	Pattern Alternatives	Performance
Water Management	ME1	**WTM:** FWS	1.20 (L) **1.20 (L)**
Waste Management+	E2 E4 NA1	**WSM:** ORC-HRU---**NM:** D-ICAOS-AS-FO **WSM:**ORC---**WTM:**FWS **WSM:**ORC-ORP	4.67 (H) 0.46 (L) 5.00 (H) **3.41 (M)**
GHG &Noise Management	E3 ME3 SA1 NA3 NA6	**GHG1:** F---**NM:** D-ICAO-FO **GHG1:** D-R-O-W-F-M---**NM:**AD **GHG1:** R-O-M---**NM:**D-ICAOS **GHG1:** VD-VO---**GHG2:** A---**NM:**FAAS **GHG2:** A-RF---**NM:**D	3.32(M) 5.00 (H) 3.63 (H) 1.44 (L) 0.23 (L) **2.73 (M)**
GHG Emissions	E5 A1	**GHG1:** R-O-W---**GHG2:** SE-RF **GHG2:** ES-A-SE	2.21 (M) 1.80 (M) **2.00 (M)**
GHG, Waste & Noise	A5 NA2 NA5	**GHG1**:O---**WSM:**ORC---**NM:**D **GHG2:** ES-A---**WSM:** ORC-ORP- IUP-IRU---**NM:** ICAOS **GHG1:** R-F---**WSM:** ORC-OUP-IRC-IUP-IRU-HRU---**NM:**D-ICAOS	0.79 (L) 2.83 (M) 3.17 (M) **2.26 (M)**
GHG & Water Management+	A4 A7	**GHG2:** ES---**WTM:**MWR-FWR-FWS **GHG1:**D-R-M---**WTM:**FWR---**NM:**ICAOS	2.67 (M) 3.38 (H) **3.03 (M)**
Do All	E1	**GHG1:** O-F- VD-VM ---**GHG2:**ES-SE-RF---**WSM:** OUP---**WTM:** MWR-MWS-FWS---**NM:** D-ICAOS-FO	**0.90 (L)**
Noise Management	A2 A3 A6 ME2	**NM:** D-FO **NM:**ICOS **NM:**FO **NM:**ICAOS	---- ---- ---- ----

(L): Low **(M):** Moderate **(H):** High
Note: to have a better insight about the strategy patterns' acronyms, please see Table 2.

DISCUSSION

The number of taxonomies reported by this study was eight. This number is more than the number of reported patterns by the previous studies. The number of patterns have been reported by the previous studies was four patterns (e.g. Migdadi & Omari, 2019; Migdadi, 2019), or three patterns (e.g. Migdadi, 2016; Migdadi & Elzaqaiba, 2018). This study adopted more in-depth analysis to analyze the effective strategy patterns under each indicator by each case and form the general effective strategy patterns. On other hand, the approach of developing taxonomies that adopted by this study differs from the previous studies approach. The most of previous studies adopted K-means clustering analysis, however, this study identified the strategy patterns according to the maximum number of shared sub-patterns between cases.

Some of the reported strategy patterns by this study were very close to those reported by the previous studies. For example, waste management+ pattern was very close to the hospitals' non-hazard wastes pattern that was reported by Migdadi & Omari (2019), also close to mobile phone producers' solid waste and recycling pattern that reported by Migdadi (2016), and close to the airlines' group 3 strategy pattern that reported by Migdadi (2018). Moreover, GHG emissions strategy pattern is close to the airlines' low emphasizer strategy pattern that reported by Migdadi (2019). In addition, GHG waste & noise strategy pattern was close to the airlines' low-to-moderate emphasizer pattern that reported by Migdadi (2019). Furthermore, do all strategy pattern was very close to the manufacturing sector's agile strategy pattern that was reported by Migdadi & Elzaqaiba (2018).

Some of the strategy patterns that reported by this study were not shared with those reported by previous studies. For example, the water management, GHG & water management, and the noise strategy patterns were reported only by this study. This could be related to the methodology that was adopted by this study, which differs from those adopted by previous studies. The adopted methodology was more in-depth and depended on identifying the shared sub-strategy patterns between cases. On the other hand, some indicators as the noise indicator have not been investigated by the previous studies as Migdadi (2018, 2019). Accordingly, this study is more comprehensive and detailed in comparison with the previous studies.

The achieved performance of reported strategy patterns by this study was low, moderate and high. Most patterns achieved moderate performance. The previous studies patterns' performance was between low, moderate and high (e.g. Migdadi & Elzaqaiba, 2018, Migdadi, 2019). The achieved performance of the strategy patterns reported by this study was not the same as the shared patterns that reported by the previous studies. For example, the achieved performance of the hospitals non-hazard waste strategy pattern was high (Migdadi & Omari, 2019), also the achieved performance of mobile phone producers' solid waste and the recycling strategy pattern was high (Migdadi, 2016). However, the level of achieved performance of waste management+ that was reported by this study was moderate.

The achieved performance of airlines' low emphasizer strategy group performance was low (Migdadi, 2019). However, the achieved performance of GHG emissions strategy group that reported by this study was moderate (Migdadi & Elzaqaiba, 2018). In addition, the achieved performance of manufacturing sector's agile strategy group was high. However, the achieved performance of do all strategy group that reported by this study was low. The differences in the level of performance of the strategy patterns that reported by this study in comparison with the shared patterns of the previous studies, could be related to the differences in the sectors or the differences in the adopted methodology.

CONCLUSION

The aim of this chapter was to report the effective green operations strategy patterns of airlines. This research issue has been investigated by some studies in different sectors (hospitals, mobile phone producers, manufacturing, and airlines). The most of previous studies have adopted K-means clustering analysis to develop the strategy patterns. However, this study adopted more in-depth analysis to develop the strategy patterns. The main results of this study were: the number of green strategy patterns adopted by airlines was eight. Those patterns were: water management, waste management+, GHG & noise management, GHG emissions, GHG waste & noise, GHG & water management+, do all and noise management patterns. The level of achieved performance by these patterns was low, moderate and high. The most of reported green strategy patterns achieved a moderate performance. Some of the reported green strategy patterns by this study were close to those reported by the previous studies, but the achieved level of performance was different.

The results of this chapter will benefit the academics and the industry experts alike. The academics have more detailed reporting for the green strategy patterns of the airlines, also, they can adopt the same methodology that proposed by this chapter to develop the strategy patterns for other sectors. The experts of the industry have a clear idea about the different strategy patterns that could be adopted. Despite the significant contribution of this chapter, some recommendations could be given for the future researchers. It is recommended to investigate a larger sample size to have more valid results, it is recommended to investigate the interrelationship between the green indicators of each strategy pattern to have a better understanding for the independent, the moderating, the mediating and the dependent variables.

REFERENCES

Alkhatib, S. F. S., & Migdadi, Y. K. A. A. (2018). Airline capacity strategies: Worldwide analysis, taxonomies and regional comparison. *International Journal of Advanced Operations Management*, *10*(4), 281–303. doi:10.1504/IJAOM.2018.097261

Alkhatib, S. F. S., & Migdadi, Y. K. A. A. (2019). Taxonomies of airline scheduling strategies: Cross-regional worldwide analysis. *International Journal of Services and Operations Management*, *33*(1), 26–48. doi:10.1504/IJSOM.2019.099653

Migdadi, Y. K. (2011a). *Identifying the best practice in traditional and electronic banking operations strategies in developing economies (The case of Jordan). Balancing resources with customer expectations in rapidly developing business environment* (Doctoral dissertation). University of Bradford.

Migdadi, Y. K. A. (2012a). The developing economies' banks branches operational strategy in the era of e-banking: The case of Jordan. *Journal of Emerging Technologies in Web Intelligence*, *4*(2), 189–197. doi:10.4304/jetwi.4.2.189-197

Migdadi, Y. K. A. A. (2011b). The Impact of Adopting E-banking on Branches Operations Strategy in Developing Economies: The Case of Jordan. *Information & Communication Systems*, 208.

Migdadi, Y. K. A. A. (2012b). The order winners' operations strategy of mobile phone service providers in developing economies: The case of Jordan. *International Journal of Services and Operations Management*, *13*(3), 329–360. doi:10.1504/IJSOM.2012.049708

Migdadi, Y. K. A. A. (2013). Proposed typologies of banking operations strategy. *International Journal of Services and Operations Management*, *16*(1), 42–85. doi:10.1504/IJSOM.2013.055557

Migdadi, Y. K. A. A. (2014a). Identifying the best practices in banking operations strategy in developing economics: The case of Jordan. *International Journal of Business Excellence*, *8*(1), 38–84. doi:10.1504/IJBEX.2015.065980

Migdadi, Y. K. A. A. (2014b). The Effective Volume Flexibility Strategy of Inpatients Departments in Jordan. *International Journal of Operations and Quantitative Management*, *20*(2), 125–152.

Migdadi, Y. K. A. A. (2016). Identifying the best practices in green operations strategy of leading mobile phone producers. *International Journal of Business Excellence*, *9*(1), 92–112. doi:10.1504/IJBEX.2016.073377

Migdadi, Y. K. A. A. (2017). The effective taxonomies of mobile phone service providers' operations strategy across countries: The case of Arab Middle East Region. *Operations and Supply Chain Management*, *10*(1), 25–34. doi:10.31387/oscm0260178

Migdadi, Y. K. A. A. (2018). Identifying the best practices of airlines' green operations strategy: A cross-regional worldwide survey. *Environmental Quality Management*, *28*(1), 21–32. doi:10.1002/tqem.21575

Migdadi, Y. K. A. A. (2019). Identifying the effective taxonomies of airline green operations strategy. *Management of Environmental Quality*.

Migdadi, Y. K. A. A., & Elzzqaibeh, D. A. S. I. (2018). The evaluation of green manufacturing strategies adopted by ISO 14001 certificate holders in Jordan. *International Journal of Productivity and Quality Management*, *23*(1), 90–109. doi:10.1504/IJPQM.2018.088610

Migdadi, Y. K. A. A., & Omari, A. A. (2019). Identifying the best practices in green operations strategy of hospitals. *Benchmarking: An International Journal*, *26*(4), 1106–1131. doi:10.1108/BIJ-09-2017-0242

APPENDIX

How to Rate the Performance of Green Strategies

Table 4. Example on How to Rate the Performance of Green Strategies Adopted by Each Case

Cases	Strategy pattern title	Strategy Pattern	Relative average % of change in Direct GHG emissions	Step 2 & 3: s(Achieved performance /best performance)X 5	Step 4: Level of performance
ME3	Design +	D-R-O-W-F-M	-37.7%	(-37.7%/37.7%)X5= 5	H
SA1	Route +	R-O-M	-27.4%	(-27.4%/37.7%)X5= 3.6	H
E5	Rout +	R-O-W	-26.3%	(-26.3%/37.7%)X5= 3.5	H
NA5	Rout +	R-F	-25.5%	(-25.5%/37.7%)X5= 3.4	M
E3	Fuel	F	-25%	(-25%/37.7%)X5= 3.3	M
A1	Design + (ground operations)	VD-VM	-19.8%	(-19.8%/37.7%)X5= 2.6	M
A7	Design +	D-R-M	-13.3%	(-13.3%/37.7%)X5= 1.8	M
NA3	Design + (ground operations)	VD-VO	-11%	(-11%/37.7%)X5= 1.5	M
A5	Operating +	O	-5.4%	(-5.4%/37.7%)X5= 0.7	L
E1	Operating +	O-F	-2.4%	(-2.4%/37.7%)X5= 0.3	L
Step 1: Best Performance			-37.7%		

How to Identify the Green Operations Strategic Patterns

Table 5. Example on Step 5: How to Rate the Performance of Green Strategies Adopted by Each Case

	Direct GHG1 Strategy Pattern	Indirect GHG2 Strategy Pattern	Waste management strategy pattern	Water Management Strategy Pattern	Noise management strategy patter	Average performance
Patterns of Case (NA5)	R-F		ORC-OUP-IRC-IUP-IRU-HRU		D-ICAOS	
Rate Performance	3.4		2.95		---	3.17

Table 6. Example on Step1: How to Identify the Effective Airlines Green Strategy Patterns Adopted by Each Case

Case ID	Direct GHG emission Strategic pattern (GHG1)	Indirect GHG emission strategic patterns (GHG2)	Waste Management Strategic Patterns (WSM)	Water Management Strategic Patterns (WTM)	Noise Management Strategic Pattern (NM)	Step 1: Strategy Pattern Pattern Actions	Performance of Pattern Rate (Level)
NA1			On-boardand Ground ORC-ORDP			**WSM** WSM:ORC-ORDP	5.00 (H)
ME3	Design+ D-R-O-W-F-M				Aircraft design AD	**GHG1-NM** **GHG1:** D-R-O-W-F-M---**NM:**AD	5.00 (H)
E2			Hybrid ORC-HRU		Hybrid 1: D-ICAOS-AS-FO	**WSM-NM** **WSM:** ORC-HRU---**NM:** D-ICAOS-AS-FO	4.67 (H)
SA1	Rout+ R-O-M				Hybrid 5 D-ICAOS	**GHG1-NM** **GHG1:** R-O-M---**NM:**D-ICAOS	3.63 (H)
E5	Route+ R-O-W	Sustainable Energy+ SE-RF				**GHG1-GHG2** **GHG1:** R-O-W---**GHG2:** SE-RF	3.49 (H)
Case ID	**Direct GHG emission Strategic pattern (GHG1)**	**Indirect GHG emission strategic patterns (GHG2)**	**Waste Management Strategic Patterns (WSM)**	**Water Management Strategic Patterns (WTM)**	**Noise Management Strategic Pattern (NM)**	**Strategy Pattern Pattern Actions**	**Performance of Pattern Rate (Level)**
A7	Design+ D-R-M			Facilities and Buildings Water Recovery FWR	Conformance with ICAO Standards ICAOS	**GHG1-WTM-NM** **GHG1:**D-R-M---**WTM:**FWR---**NM:**ICAOS	3.38 (H)
NA5	Route+ R-F		Hybrid ORC-OUP-IRC-IUP-IRU-HRU		Hybrid 5 D-ICAOS	**GHG1-WSM-NM** **GHG1:** R-F---**WSM:** ORC-OUP-IRC-IUP-IRU-HRU---**NM:**D-ICAOS	3.17 (M)
A4		Energy Saving+ ES		Hybrid MWR-FWR-FWS		**GHG2-WTM** **GHG2:** ES---**WTM:**MWR-FWR-FWS	2.97 (M)
NA2		Energy Saving+ ES-A	Hybrid ORC-ORDP- IUP-IRU		Conformance with ICAO Standards ICAOS	**GHG2-WSM-NM** **GHG2:** ES-A---**WSM:** ORC-ORDP- IUP-IRU---**NM:** ICAOS	2.83 (M)
E3	Fuel F				Hybrid 2: D-ICAOS-FO	**GHG1-NM** **GHG1:** F----**NM:** D-ICAO-FO	2.41 (M)
A1	Design+ VD-VM	Energy Saving+ ES-A-SE				**GHG1-GHG2** **GHG1:** VD-VM---**GHG2:** ES-A-SE	1.80 (M)
NA3	Design+ VD-VO	Accreditation+ A			Conformance with FAAS Standards FAAS	**GHG1- GHG2-NM** **GHG1:** VD-VO---**GHG2:** A---**NM:**FAAS	1.43 (L)
Case ID	**Direct GHG emission Strategic pattern (GHG1)**	**Indirect GHG emission strategic patterns (GHG2)**	**Waste Management Strategic Patterns (WSM)**	**Water Management Strategic Patterns (WTM)**	**Noise Management Strategic Pattern (NM)**	**Strategy Pattern Pattern Actions**	**Performance of Pattern Rate (Level)**
ME1				Facilities and Buildings Water Saving FWS	Flights operating FO	**WTM-NM** **WTM:**FWS---**NM:** FP	1.20 (L)

continued on following page

Table 6. Continued

A5	Operating+ O		On-boardand Ground Wastes ORC		Aircraft design D	**GHG1-WSM-NM** **GHG1**:O---**WSM**:ORC---**NM**:D	0.79 (L)
E1	Operating+ O-F	Energy Saving+ ES-SE-RF	On-boardand Ground OUP	Hybrid MWR-MWS-FWS	Hybrid 2: D-ICAOS-FO	**GHG1-GHG2-WSM-WTM-NM** **GHG1**: O-F---**GHG2**:ES-SE-RF---**WSM**: OUP---**WTM**: MWR-MWS-FWS---**NM**: D-ICAOS-FO	0.62 (L)
E4			On-boardand ground ORC	Facilities and Buildings Water FWS		**WSM-WTM** **WSM**:ORC---**WTM**:FWS	0.43 (L)
NA6		Accreditation+ A-RF			Aircraft design D	**GHG2-NM** **GHG2**: A-RF---**NM**:D	0.23 (L)
Case ID	**Direct GHG emission Strategic pattern (GHG1)**	**Indirect GHG emission strategic patterns (GHG2)**	**Waste Management Strategic Patterns (WSM)**	**Water Management Strategic Patterns (WTM)**	**Noise Management Strategic Pattern (NM)**	**Strategy Pattern** **Pattern Actions**	**Performance of Pattern Rate (Level)**
ME2					Conformance with ICAO Standards ICAOS	**NM** **NM**: ICAOS	----
A3					Conformance with ICAO Standards ICAOS	**NM** **NM**: ICAOS	----
E6					Hybrid 4: D-AS	**NM** **NM**:D-AS	----
A2					Hybrid 3: Design-Operating D-FO	**NM** **NM**: D-FO	------
A6					Flights operating FO	**NM** **NM**: FO	----

Table 7. Example on How to Identify Airline Effective Green Operations Strategy Patterns

Case	Step 1: Green Operations Pattern	Pattern rated performance
NA5	**GHG1-WSM-NM** **GHG1**: R-F---**WSM**: ORC-OUP-IRC-IUP-IRU-HRU---**NM**:D-ICAOS	3.17 (M)
NA2	**GHG2-WSM-NM** **GHG2**: ES-A---**WSM**: ORC-ORDP- IUP-IRU---**NM**: ICAOS	2.83 (M)
A5	**GHG1-WSM-NM** **GHG1**:O---**WSM**:ORC---**NM**:D	0.79 (L)
	Step 2: General strategy pattern **GHG-WSM-NM**	**Step 3:** Average performance for the general pattern (3.17+2.83+0.79)/3= 2.26 (M)

Chapter 8
Airline Effective Green Operations Strategy Patterns:
Regional Level Analysis

ABSTRACT

This chapter reports airline effective green operations strategy patterns adopted by each region. To achieve this aim, green practices of 23 airlines from five regions were investigated. The data used in this chapter was the effective green operations strategy that adopted by each airline, which is the result of the previous chapter. The chapter reported the effective green strategy patterns that adopted by each region. The effectiveness of these strategy patterns was moderate in Asia, Europe, the Middle East, and North America, but high in South America. This chapter helps decision makers and academics alike, since the decision makers can adopt the most effective patterns. In addition, the academics have in-depth insight about the green strategy patterns that adopted by each region, so the propositions and hypotheses of future researches could be formulated according to the results of this study.

INTRODUCTION

Green operations strategy and the practices of airlines differ across regions (Migdadi, 2020). Since the environmental effect of civil aviation differs across regions. For example, the emissions of Australian civil aviation sector in 2010 was 165% above the level of 1990 (Bureau of Infrastructure

DOI: 10.4018/978-1-7998-4255-2.ch008

Transport & Regional Economics, 2009). CO2 emission of the aviation sector in European region reached 163 million tons in 2017, by an increase of 16% in comparison with 2015 (European Union Aviation Safety Agency, 2019). Such differences in the environmental effect reflects the differences in the region traffic. For example, the largest region in activity was Asia with 35% of the world traffic, followed by Europe and North America with 27% for each region (Gittens, et al., 2017).

In this regard, the airlines in different regions should take more effective actions to reduce the environmental effect of their operations. Accordingly, the airlines should have a better insight about the most effective strategy patterns that should be adopted. Airline effective green operations strategies of each region have been reported by very limited studies (e.g. Migdadi, 2018, 2019, 2020). Some of these studies have reported the airline membership for a particular green strategy pattern, with limited concern about investigating in-depth the strategy patterns that adopted by each region. In addition, the previous studies reported the patterns according to the general action categories, with limited concern about the detailed actions taken and its impact (e.g. Migdadi, 2020).

This chapter bridged the gap of previous studies by reporting the effective green operations strategy patterns that adopted by each region's airlines. This study reported the practices of many airlines in different regions, and developed the effective strategic patterns. This chapter depended on the results of previous chapter. Accordingly, the main question of this chapter was: what were the most effective green strategic patterns that adopted by each region's airlines? To answer the question the following objectives were realized:

1. Identifying the green operations strategy patterns the adopted by airlines of each region.
2. Evaluating the effectiveness of strategy patterns that adopted by each airline of each region.
3. Identifying the effectiveness of strategy patterns that adopted by each region.

This chapter is arranged in sections: the first one is the literature review, followed by the methodology and methods, then, the results of the empirical study, next, the discussion, and the final section is the conclusions.

LITERATURE REVIEW

Studying the differences in the airline operations strategy across regions have been investigated by some limited studies (e.g. Alkhatib & Migdadi, 2018a; Alkhatib & Migdadi, 2018b, Alkhatib & Migdadi, 2019a, Alkhatib & Migdadi, 2019b). Different capacity strategy patterns adopted by the airlines in different regions, for example, 50.8% of Asia region's airlines adopted very low capacity. However, 42.8% of Africa region's airlines adopted low capacity-moderate destination pattern, and same percentage of Africa region's airlines adopted very low capacity strategy (Alkhatib & Migdadi, 2018a). Moreover, scheduling strategies adopted by Arab states region were: focused international, focused regional, and focused intercontinental. However, the Asia and Africa region's airlines adopted two scheduling patterns: focused regional and focused international. North America and South America region's airlines adopted regional focused strategy (Alkhatib & Migdadi, 2019a).

The operational determinants of airline service quality differs across regions. The perceived quality of Arab states region's airline was affected by the number of aircrafts, and the number of destinations. The perceived quality of North America region's airlines was affected by the number of hubs. However, the quality of Asia airlines was affected by the number of seats. The quality of South America airlines was affected by the number of destinations (Alkhatib & Migdadi, 2018b). On the other hand, the strategic alignment between the quality and the scheduling strategies affects the quality of airlines. Different alignment patterns were adopted by the airlines in the different regions. For example, eight alignment patterns were adopted by Africa region's airlines; half of the patterns were not effective. Thirteen alignment patterns were adopted by Arab States region's airlines, fourteen alignment patterns were adopted by Asia region's airlines, and twelve alignment patterns were adopted by North America region's airlines (Alkhatib & Migdadi, 2019a).

Limited research contribution have investigated the green operations strategy that adopted by different regions (e.g. Migdadi, 2018, Migdadi & Omari, 2019, Migdadi, 2019,). Some of the previous studies have reported in-depth the strategy patterns adopted by each region (e.g. Migdadi, 2018, 2019). The investigation was too general, as the regional membership for each strategy patterns, the detailed strategy patterns that adopted by each region, and the effectiveness of these patterns have not been investigated by the previous studies (e.g. Migdadi, 2018, 2019).

The effective green operations strategy that adopted by airlines differs across regions. According to Migdadi (2019), the effective taxonomies of green operations strategy that adopted by Asia region's airlines were the low-to-moderate emphasizer strategy pattern, and the low emphasizer strategy patterns, each pattern was adopted by two airlines. One airline in Europe region adopted the high emphasizer, and two airlines adopted the low-to-moderate emphasizer. Only one pattern was adopted by one airline in North America region, this pattern was the low emphasizer. No any effective strategy pattern was adopted by the South America region's airlines. One pattern was adopted by one airline in the Middle East, which was the high emphasizer. The same close result was reached by Migdadi (2020). The airlines of Middle East region achieved the highest green index, this index was about 91%, followed by the Europe region's airlines, which was 75%. The next region was South America region, the achieved green index was 52%, followed by Asia region's airlines, which achieved an index of 43%. The lowest green index was achieved by North America region's airlines, which was about 14%.

According to another study of Migdadi (2020), the effective green strategy of Asia airlines achieved a significant performance in GHG2, the energy saving, on-board and the ground wastes upcycling, and upcycling of the hazard wastes. The effective strategy of the Middle East airlines achieved a significant performance in GHG2, energy, and saving water in facilities and buildings. North America effective strategy achieved significant reduction in GHG emissions. However, Europe region's airlines achieved a significant performance in the wastes recycling, the water consumption and the noise minimization. South America airlines achieved the significant reduction in GHG1 emissions and the fuel consumption. Migdadi (2018) study reached very close results. According to this study, three strategies were adopted by Europe region's airlines; the most adopted pattern was GHG, resource management & waste management. In addition, three patterns were adopted by North America region's airlines, the most adopted pattern was GHG, resources management & waste management.

Four patterns were adopted by Asia airlines, the most adopted patterns were GHG, resources management & waste management. The second pattern was GHG & resources management. Each pattern was adopted by two airlines in Asia. Two patterns were adopted by the Middle East region's airlines. The adopted patterns were GHG & resources management, and water management & resources management. Only one strategy pattern was adopted by South America region's airlines, which was GHG & resources management. According to the previous studies of the airlines green operations strategy, the most effective region in the green performance was the Middle East, followed by Europe and the lowest effective region was North America. In addition, the widely shared strategy patterns across regions was GHG & resources management, GHG & energy and fuel saving. The lowest adopted strategy pattern was water management pattern, water and waste & noise management.

METHODOLOGY AND METHODS

This part of the chapter discussed how the strategic patterns were identified, and how the effectiveness of these patterns were assessed. In this regard, the design of sample was discussed, also, how the data was collected, and how the data was analyzed to formulate the strategy patterns, and track the effectiveness of these patterns. In addition, supportive examples were presented in the appendices.

Regions Classification

According to the previous studies, the regions of the world are classified into five regions; Africa, Asia & Pacific, Europe, Latin America (South America), the Middle East and North America (Gittens et al., 2017; Herdman, 2018; Migdadi, 2018; Migdadi, 2019; RobecoSAM, 2019; Migdadi, 2020). Accordingly, this chapter adopted the same classification scheme.

Sample Design

The unit of analysis for this chapter is the region. Accordingly, 23 airlines which adopted the effective green operations strategy of five regions were surveyed. The sample size is a purposive accordingly to the adopting of effective green operations strategy. The distribution of sample across regions was as presented in Table 1. It can be seen that seven cases were surveyed in Asian, seven cases were surveyed in Europe, however, five cases were surveyed in North America, three cases in the Middle East and one case in South America.

Sources of Data

This chapter depended on the analysis outcomes of the previous chapter. Accordingly, the data required for this chapter was the adopted effective strategy patterns to achieve the green performance related to each green performance indicator.

Table 1. Sample of study

Regions	Samples size
A: Asia	7
E: Europe	7
NA: North America	5
ME: Middle East	3
SA: South America	1
Total	23

Data Analysis

The purpose of data analysis was to assess the performance of each green strategy pattern that adopted by each region. Also, to assess the overall green performance of each region's strategy patterns. Accordingly, the following steps were taken to achieve this purpose:

Step 1: Identifying the effective strategy patterns adopted by the airlines (this is the outcome of the previous chapter).

Step 2: Identifying the region of each case adopted an effective green strategy pattern. The acronym of each case represents the region initials. For example, A: Asia, E: Europe, ME: Middle East.

Step 3: Identifying the strategy patterns adopted by each region according to the cases membership for each strategy pattern.

Step 4: Identifying the performance of each alternative under each strategy pattern.

Step 5: Computing the average rated performance of all cases under each strategy pattern, to find the pattern average performance

Step 6: Computing the average performance achieved by the effective patterns of each region by computing the average of performance of the strategy alternatives.

Please see the example in the Appendix, Tables 9 through 11.

RESEARCH RESULTS

This part of the chapter presented the main results of the empirical study. The aim of this part is to identify the effective green operations strategies that adopted by each region. This part started by presenting the summary of airline green operations strategies that adopted by each region. Then, the details of the strategic patterns adopted by each region were presented. In order to facilitate the reading of the strategy patterns alternatives, the following table (Table 2) shows the acronyms of the strategy patterns alternatives.

Airline Effective Green Operations Strategy across Regions

Table 3 shows the summary of airline effective green strategy patterns that adopted by each region. It can be seen that each region adopted a different combination and different number of effective green operations strategy. For example, Asia adopted four patterns: GHG emissions, GHG, waste & noise, GHG & water management+, and noise management. Europe region adopted four patterns: waste management+, GHG & noise management, GHG emissions, and do all. The Middle East region adopted three patterns: water management, GHG & noise management, and noise management. North America adopted three patterns: waste management+, GHG & noise management, GHG, waste & noise. South America adopted one pattern, GHG & Noise Management.

The widely shared pattern between the regions was GHG & noise management. Four regions, Europe, the Middle East, North America, and South America adopted this pattern. Different performance achieved by this pattern across regions, it was high by the Middle East and South America, moderate by Europe and low by North America. The range of strategy patterns performance differs, all patterns of Asia achieved moderate performance. The performance achieved by Europe, the Middle East, and North America

Table 2. List of Acronyms Used in Presenting the Chapter Results

Acronym	Interpretation
A	Facilities and building accreditation
AS	Conformance with Airport fly quiet program
D	Aircraft Design
ES	Facilities and building energy saving
F	Flights fuel management
FAAS	Aircraft conformance with FAA noise standards
FO	Flights operating
FWR	Facilities and buildings water recycling and recovery
FWS	Facilities and buildings water saving
GHG1	Direct greenhouses gas emission
GHG2	Indirect greenhouse gas emission
HRU	Hazard wastes reusing
HUP	Hazard wastes upcycling
HRC	Hazard wastes recycling
ICAOS	Aircraft conformance with ICAO Noise Standards
IRC	Industrial wastes recycling
IRU	Industrial wastes reusing
IUP	Industrial wastes upcycling
M	Flights maintenance management
MWR	Maintenance water recycling and recovery
MWS	Maintenance water saving
NM	Noise management
O	Flight Operating
ORC	On-board and ground wastes recycling
ORP	On-board and ground reduce using of paper
OUP	On-board and ground wastes upcycling
R	Flights routes management
RF	Upgrading and replacing facilities
SE	Facilities and building sustainable energy
VD	Engine and vehicle design
VO	Engine and vehicle Operating
VM	Engine and vehicle maintenance
W	Flights weight management
WSM	Waste management
WTM	Water Management

Table 3. Airline Effective Green Strategy Patterns Adopted by Each Region

Strategy Pattern	Asia	Europe	The Middle East	North America	South America
Water Management			1.20 (L)		
Waste Management+		2.61 (M)		5.00 (H)	
GHG &Noise Management		2.41 (M)	5.00 (H)	0.84 (L)	3.63 (H)
GHG Emissions	1.80 (M)	2.51 (M)			
GHG, Waste & Noise	0.79 (L)			3 (M)	
GHG & Water Management+	3.03 (M)				
Do All		0.90 (L)			
Noise Management	NRP		NRP		
Airline Region Performance	1.87 (M)	2.21 (M)	3.10 (M)	2.53 (M)	3.63 (H)

(L): Low **(M):** Moderate **(H):** High
NRP: No Reported Performance

was low, moderate, and high. However, the achieved performance by South America was high. The overall region performance was moderate for four regions, Asia, Europe, the Middle East and North America. In addition, the performance achieved was high by South America.

Airline Effective Green Operations Strategy of Asia Region

Table 4 shows the strategy patterns that adopted by the Asia region airlines. It can be seen that the adopted patterns were four: GHG emissions, GHG waste & noise, GHG & water management+, and noise management. The achieved performance by these strategy patterns differs. For example, two patterns achieved a moderate performance, which were GHG Emissions, and GHG & water management+. The number of alternatives of each strategy pattern was one for GHG emissions, GHG, waste & noise, and GHG & water management+. However, the number of noise management pattern alternatives was three. The overall achieved performance by the effective patterns of this region was moderate.

Table 4. Airline Effective Green Strategy Patterns adopted by Asia Region

Strategy Pattern	Pattern Alternatives	Performance
GHG Emissions	**GHG2:** ES-A-SE	1.80 (M)
GHG, Waste & Noise	**GHG1**:O---**WSM:**ORC---**NM:**D	0.79(L)
GHG & Water Management+	**GHG2:** ES---**WTM:**MWR-FWR-FWS	3.03 (M)
Noise Management	**NM:** D-FO **NM:**ICOS NM:FO	NRP
Region Effective Strategy Patterns' Performance		**1.87 (M)**

(L): Low **(M):** Moderate **(H):** High
Note: to have a better insight about the strategy patterns' acronyms, please see Table 2.

Airline Effective Green Operations Strategy of Europe Region

Table 5 shows the strategy patterns adopted by Asia region. It can be seen that the adopted patterns were four: waste management+, GHG & noise management, GHG emissions and do all. The achieved performance by these strategy patterns was moderate by three patterns, which were waste management+, GHG & noise management, and GHG emissions. However, one pattern which was do all achieved a low performance. The number of alternatives of each strategy pattern was two for waste management+, and one pattern for GHG & noise management, GHG emissions and do all. The overall performance achieved by the effective patterns of this region was moderate.

Table 5. Airline Effective Green Strategy Patterns adopted by Europe Region

Strategy Pattern	Pattern Alternatives	Performance
Waste Management+	**WSM:** ORC-HRU---**NM:** D-ICAOS-AS-FO **WSM:**ORC---**WTM:**FWS	4.76 (H) 0.46 (L) **2.61 (M)**
GHG &Noise Management	**GHG1:** F---**GHG2:** SE-RF---**NM:** D-ICAO-FO	2.41 (M)
GHG Emissions	**GHG1:** R-O-W---**GHG2:** SE-RF	2.51 (M)
Do All	**GHG1:** O-F-VD-VM ---**GHG2:**ES-SE-RF---**WSM:** OUP---**WTM:** MWR-MWS-FWS---**NM:** D-ICAOS-FO	0.90 (L)
Region Effective Strategy Pattern's Performance		**2.21 (M)**

(L): Low **(M):** Moderate **(H):** High
Note: to have a better insight about the strategy patterns' acronyms, please see Table 2.

Airline Effective Green Operations Strategy of the Middle East Region

Table 6 shows the strategy patterns adopted by the Middle East region. It can be seen that the adopted patterns were three: water management, GHG & noise management and noise management. The achieved performance by water management pattern was low. However, the performance of GHG & noise management pattern was high. The number of alternatives of each strategy pattern was one for all patterns. The overall performance achieved by the effective patterns of this region was moderate.

Table 6. Airline Effective Green Strategy Patterns adopted by the Middle East Region

Strategy Pattern	Pattern Alternatives	Performance
Water Management+	**WTM:** FWS	1.20 (L)
GHG &Noise Management	**GHG1:** D-R-O-W-F-M---**NM:**AD	5.00 (H)
Noise Management	**NM:**ICAOS	NRP
Region Effective Strategy Patterns Performance		**3.10 (M)**

(L): Low **(M):** Moderate **(H):** High
Note: to have a better insight about the strategy patterns' acronyms, please see Table 2.

Airline Effective Green Operations Strategy of North America Region

Table 7 shows the strategy patterns adopted by North America region. It can be seen that the adopted patterns were three: waste management+, GHG & noise management, GHG, waste & noise. The achieved performance by these strategy patterns differs. For example, the achieved performance was high by water management+, and low by GHG, Waste & Noise. The number of alternatives of each strategy pattern was two for GHG & noise management, and GHG, Waste & Noise. One pattern was adopted by waste management+. The overall performance achieved by the effective patterns of this region was moderate.

Table 7. Airline Effective Green Strategy Patterns adopted by North America Region

Strategy Pattern	Pattern Alternatives	Performance
Waste Management+	**WSM:** ORC-ORP	5.00 (H)
GHG &Noise Management	**GHG1:** VD-VO---**GHG2:** A---**NM:**FAAS **GHG2:** A-RF---**NM:**D	1.44 (L) 0.23 (L) **0.84 (L)**
GHG, Waste & Noise	**GHG2:** ES-A---**WSM:** ORC-ORP- IUP-IRU---**NM:** ICAOS **GHG1:** R-F---**WSM:** ORC-OUP-IRC-IUP-IRU-HRU---**NM:**D-ICAOS	2.83 (M) 3.17 (M) **3.00 (M)**
Region Effective Strategy Patterns Performance		**2.53 (M)**

(L): Low **(M):** Moderate **(H):** High
Note: to have a better insight about the strategy patterns' acronyms, please see Table 2.

Airline Effective Green Operations Strategy of South America Region

Table 8 shows the strategy patterns adopted by South America region. It can be seen that the adopted patterns was one: GHG & Noise Management. The achieved performance by the strategy pattern was high. The number of alternatives of this strategy pattern was one. The overall performance achieved by the effective patterns of this region was high.

Table 8. Airline Effective Green Strategy Patterns adopted by South America Region

Strategy Pattern	Pattern Alternatives	Performance
GHG &Noise Management	**GHG1:** R-O-M---**NM:**D-ICAOS	3.63 (H)
Region Effective Strategy Patterns Performance		**3.63 (H)**

(L): Low **(M):** Moderate **(H):** High
Note: to have a better insight about the strategy patterns' acronyms, please see Table 2.

DISCUSSION

Each region adopted different strategy pattern in comparison with other regions. This is the same result reached by Migdadi (2018, 2019, 2020). This reflects the differences in the operational characteristics as the capacity and the flights scheduling as reported by Alkhtib & Migdadi (2018a, 2019a).

Also, reflect the differences in the region traffic. The widely adopted pattern across regions was GHG & noise management. This pattern was adopted by four regions: Europe, the Middle East, North America and South America. The frequency of aircrafts movements and hence the noise event has generally increased (Glasgow Airport, 2014), the noise tends to cover larger geographical areas, and the noise is more difficult to mitigate when it is compared to road and rail noise (Civil Aviation Authority, 2014). In addition, the concern about reducing GHG emissions is also widely adopted across regions since the environmental impact of airlines has increased over the last 25 years as a result of the increase in demand (Lee *et al.,* 2009). CO2 emissions of airlines are expected to increase by further 45% between 2014 and 2035. The future green targets of aviation as adopted by IATA is improving the average efficiency in fuel by 1.5% per year from 2009 to 2020, and reducing in the net aviation CO2 emissions of 50% by 2050 relative to 2005 level (International Air Transport Association, 2019).

The effective strategy patterns performance was moderate for four regions: Asia, Europe, the Middle East and North America. In addition, the performance achieved was high by South America region. This result is very close to the results reached by Migdadi (2019) study, accordingly to this study, Asia airlines adopted two patterns, which were: the low-to-moderate emphasizer strategy pattern, and the low-emphasizer strategy patterns, each pattern was adopted by two airlines. One airline in Europe region adopted the high emphasizer, and two airlines adopted the low-to-moderate emphasizer. Only one pattern was adopted by one airline in North America, this pattern was the low emphasizer. This differences in the performance of adopted strategy patterns by each region in comparison with others reflects the differences in traffic. For example, the airline sector is the largest in Asia region with 33% market share, followed by Europe and North America with 27%. The Middle East region represents 9% of the world traffic, Latin America and Caribbean region accounted for 5%, while the remaining 2% of the world traffic was undertaken by African region (Gittens *et al.,* 2017).

CONCLUSION

The aim of this chapter was to report the airline effective green operations strategies that adopted by each regions. This research issue has been raised by limited studies, most of the previous studies have reported the airline membership for a particular green strategy pattern, with limited concern about

Table 9. **Step 1:** *How to Identifying the Airline Effective Green Operations Strategy Patterns (Results of the Previous Chapter)*

Pattern	Case ID	Pattern Alternatives	Performance
Water Management	ME1	**WTM:** FWS	1.20 (L) **1.20 (L)**
Waste Management+	E2 E4 NA1	**WSM:** ORC-HRU---**NM:** D-ICAOS-AS-FO **WSM:**ORC---**WTM:**FWS **WSM:**ORC-ORP	4.76 (H) 0.46 (L) 5.00 (H) **3.41 (M)**
GHG &Noise Management	E3 ME3 SA1 NA3 NA6	**GHG1:** F---**NM:** D-ICAO-FO **GHG1:** D-R-O-W-F-M---**NM:**AD **GHG1:** R-O-M---**NM:**D-ICAOS **GHG1:** VD-VO---**GHG2:** A---**NM:**FAAS **GHG2:** A-RF---**NM:**D	3.32(M) 5.00 (H) 3.63 (H) 1.44 (L) 0.23 (L) **2.73 (M)**
GHG Emissions	E5 A1	**GHG1:** R-O-W---**GHG2:** SE-RF **GHG2:** ES-A-SE	2.21 (H) 1.80 (M) **2.00 (M)**
GHG, Waste & Noise	A5 NA2 NA5	**GHG1**:O---**WSM:**ORC---**NM:**D **GHG2:** ES-A---**WSM:** ORC-ORP- IUP-IRU---**NM:** ICAOS **GHG1:** R-F---**WSM:** ORC-OUP-IRC-IUP-IRU-HRU---**NM:**D-ICAOS	0.79 (L) 2.83 (M) 3.17 (M) **2.26 (M)**
GHG & Water Management+	A4 A7	**GHG2:** ES---**WTM:**MWR-FWR-FWS **GHG1:**D-R-M---**WTM:**FWR---**NM:**ICAOS	2.67 (M) 3.38 (H) **3.03 (M)**
Do All	E1	**GHG1:** O-F- VD-VM ---**GHG2:**ES-SE-RF---**WSM:** OUP--- **WTM:** MWR-MWS-FWS---**NM:** D-ICAOS-FO	**0.90 (L)**
Noise Management	A2 A3 A6 ME2	**NM:** D-FO **NM:**ICOS **NM:**FO **NM:**ICAOS	---- ---- ---- ----

(L): Low **(M):** Moderate **(H):** High
Note: to have a better insight about the strategy patterns' acronyms, please see Table 2.

investigating in-depth the strategy patterns that adopted by each region. The main results of this study were: each regions adopted different combination and the number of effective green operations strategy. Asia region's airlines adopted four patterns: GHG emissions, GHG, waste & noise, GHG & water management+ and noise management. Europe region's airlines adopted four patterns: waste management+, GHG & noise management, GHG emissions, and do all. Middle East region's airlines adopted three patterns: water management, GHG & noise management, and noise management. North America region's airlines adopted three patterns: waste management+, GHG & noise management, and GHG, waste & noise. South America region's airlines adopted one pattern: GHG & noise management. The overall region performance was moderate for four regions' airlines: Asia, Europe, the Middle

East and North America. In addition, the performance achieved was high by South America region's airlines. The widely shared pattern between regions was GHG & noise Management. This pattern was adopted by four regions' airlines: Europe, the Middle East, North America and South America.

The results of this chapter will benefit the academics and the industry experts alike. The academics have in-depth results about the effective strategy patterns that adopted by each region. Therefore, they can develop hypotheses for the future researchers. The experts in the industry have a clear idea about the different strategy patterns that could be adopted. Despite the significant contribution of this chapter, some recommendations could be given for the future researchers. It is recommended to investigate a larger sample size to have more valid results, it is recommended to investigate the interrelationship between the green actions and indicators of each region's airlines to have a better understanding for the independent, the moderating, the mediating and the dependent variables.

APPENDIX: HOW TO IDENTIFY THE STRATEGY PATTERNS OF EACH REGION

Table 10. Identifying the Effective Strategy Patterns Adopted by Each Region According to the cases Membership for Each Strategy Pattern.

Step 3: Strategy Pattern	Case	Strategy Alternative	Step 4: Performance
Waste Management+	E2 E4	**WSM:** ORC-HRU---**NM:** D-ICAOS-AS-FO **WSM:**ORC---**WTM:**FWS **Step 5: pattern alternative performance**	4.76 (H) 0.46 (L) **(4.76+0.46)/2= 2.61**
GHG &Noise Management	E3	**GHG1:** F---**GHG2:** SE-RF---**NM:** D-ICAO-FO	2.41(M)
GHG Emissions	E5	**GHG1:** R-O-W---**GHG2:** SE-RF	2.51 (M)
Do All	E1	**GHG1:** O-F- VD-VM ---**GHG2:**ES-SE-RF---**WSM:** OUP---**WTM:** MWR-MWS-FWS---**NM:** D-ICAOS-FO	0.90 (L)

Table 11. How to Compute the Average Performance Achieved by the Effective Patterns of Each Region

Strategy Pattern	Strategy Alternative	Step 4: Performance
Waste Management+	**WSM:** ORC-HRU---**NM:** D-ICAOS-AS-FO **WSM:**ORC---**WTM:**FWS	4.76 (H) 0.46 (L)
GHG &Noise Management	**GHG1:** F---**GHG2:** SE-RF---**NM:** D-ICAO-FO	2.41(M)
GHG Emissions	**GHG1:** R-O-W	2.51 (M)
Do All	**GHG1:** O-F- VD-VM ---**GHG2:**ES-SE-RF---**WSM:** OUP---**WTM:** MWR-MWS-FWS---**NM:** D-ICAOS-FO	0.90 (L)
Step 6: Compute the Average Performance of the Effective patterns Each Region		(4.76+0.46+2.41+2.51+0.90)/5= **2.21 (M)**

REFERENCES

Alkhatib, S., & Migdadi, Y. (2019b). Operations Strategies Alignment Patterns and Service Quality: Airline Global Analysis. *The Quality Management Journal*, *26*(3), 144–156. doi:10.1080/10686967.2019.1620590

Alkhatib, S. F., & Migdadi, Y. K. (2018b). Operational determinants of airline service quality: Worldwide cross-regional analysis. *The Quality Management Journal*, *25*(4), 186–200. doi:10.1080/10686967.2018.1515525

Alkhatib, S. F. S., & Migdadi, Y. K. A. A. (2018a). Airline capacity strategies: Worldwide analysis, taxonomies and regional comparison. *International Journal of Advanced Operations Management*, *10*(4), 281–303. doi:10.1504/IJAOM.2018.097261

Alkhatib, S. F. S., & Migdadi, Y. K. A. A. (2019a). Taxonomies of airline scheduling strategies: Cross-regional worldwide analysis. *International Journal of Services and Operations Management*, *33*(1), 26–48. doi:10.1504/IJSOM.2019.099653

Bureau of Infrastructure, Transport and Regional Economics. (2009). *Greenhouse gas emissions from Australian transport: projections to 2020.* Working paper 73. https://www.bitre.gov.au/publications/2009/files/wp_073.pdf

Civil Aviation Authority. (2014). *Managing aviation noise*. https://publicapps.caa.co.uk/docs/33/CAP%201165%20Managing%20Aviation%20Noise%202.pdf

European Union Aviation Safety Agency. (2019). European Aviation. *Environment Reporter, 2019*. https://www.eurocontrol.int/sites/default/files/publication/files/eaer-2019.pdf

Gittens, A., Juniac, A., Pie, J., Poole, J., & Liu, F. (2017). *Aviation Benefits 2017*. ICAO. https://www.icao.int/sustainability/Documents/AVIATION-BENEFITS-2017-web.pdf

Glasgow Airport. (2014). *Noise action plan 2013-2018*. https://www.glasgowairport.com/media/2067/current-noise-action-plan.pdf

Herdman, D. (2018). *Asia Pacific Aviation: Competitive dynamics & growth.* Willis Towers Watson Aviation Conference, Singapore.

International Air Transport Association. (2019). *Improving environmental performance*. https://www.iata.org/whatwedo/environment/Pages/index.aspx

Lee, D. S., Fahey, D. W., Forster, P. M., Newton, P. J., Wit, R. C., Lim, L. L., ... Sausen, R. (2009). Aviation and global climate change in the 21st century. *Atmospheric Environment*, *43*(22-23), 3520–3537. doi:10.1016/j.atmosenv.2009.04.024

Migdadi, Y. K. A. A. (2018). Identifying the best practices of airlines' green operations strategy: A cross-regional worldwide survey. *Environmental Quality Management*, *28*(1), 21–32. doi:10.1002/tqem.21575

Migdadi, Y. K. A. A. (2019). Identifying the effective taxonomies of airline green operations strategy. *Management of Environmental Quality*.

Migdadi, Y. K. A. A. (2020). Green Operational Strategy for Airlines: Content and Regional Analysis. In Cases on Green Energy and Sustainable Development (pp. 193-229). IGI Global.

Migdadi, Y. K. A. A., & Omari, A. A. (2019). Identifying the best practices in green operations strategy of hospitals. *Benchmarking: An International Journal*, *26*(4), 1106–1131. doi:10.1108/BIJ-09-2017-0242

Robeco, S. A. M. (2018). *DJSI 2018 Review Results* September 2018. https://www.robecosam.com/media/0/4/3/043cd016c8d1952fdd23ee4b2dac909d_review-presentation-2018_tcm1016-14658.pdf

Chapter 9
Airline Green Operations Strategy:
Future Recommendations

ABSTRACT

This chapter recommends future improvement on airline green operations strategy patterns. To achieve this aim, the ineffective green operations strategy of 12 airlines from five regions were analyzed, and changes on the ineffective strategies have been recommended. The average overall index of all green indicators will be improved to be positive. All indices of all regions will be improved to be positive. The number of effective patterns after improvement will be seven. In addition, the green indices achieved after improvement will be moderate. This chapter helps the decision makers and the academics alike. The academics have in-depth results about the most effective strategy patterns that could be utilized to develop more extended theoretical models. The experts in the industry have a clear idea about the recommended effective strategy patterns that could be adopted.

INTRODUCTION

The focus of this book was on reporting the effective green operations strategy that adopted by airlines. However, it was found during the investigation that some airlines have adopted green operations strategy but these strategies were ineffective. The previous studies of green operations strategies have found

DOI: 10.4018/978-1-7998-4255-2.ch009

the same result. For example, Migdadi (2019) reported the misleading green operations strategy pattern, this pattern was ineffective and was adopted by 13 airlines. The most of previous studies have reported the effective green operations strategies, but these studies have not concern about recommending future changes on the current adopted strategies to have more effective strategy patterns (e.g. Migdadi, 2016; Migdadi & Elzaqaiba, 2018; Migdadi, 2018; Migdadi & Omari, 2019; Migdadi, 2019; Migdadi, 2020). Accordingly, this chapter will make more contribution in comparison with the previous studies by recommending the changes on the ineffective green strategy patterns to have more effective strategy patterns in the future. Accordingly, the main question of this chapter was: what are the recommended changes on the current ineffective green strategic patterns that adopted by airlines to have more effective strategy patterns?. To answer the question the following objectives were realized:

1. Identifying the airlines that adopted ineffective green operations strategy.
2. Evaluating the green performance of airlines and regions that adopted the ineffective green operations strategy.
3. Recommending changes on green operations strategy under each green indicator for the ineffective cases.
4. Evaluating the impact of adopting the recommended changes of green operations strategy on the green index of airlines and regions.
5. Tracing the impact of recommended changes on the green operations strategy patterns.

This chapter is arranged in sections: the first one is the methodology and methods, then, the results of the empirical study, and the final section is the conclusions.

METHODOLOGY AND METHODS

This part of the chapter discussed how to improve the ineffective strategy patterns that adopted by some cases. In this regards, the green index as an outcome of effective and ineffective strategy patterns was computed. Then, recommending effective strategy patterns for the cases that adopted the ineffective patterns. The recommended strategy patterns were for each case under each green indicator. Next, computing the green index for the cases after adopting the recommended effective green operations strategy. Finally, identifying the changes in the general green operations strategy after adopting the effective patterns. In the following sections, there is a detailed discussion for the procedures followed in this regard.

Identifying the Cases Adopted Ineffective Strategy Patterns

The ineffective strategy patterns are those adopted by some airlines but achieved a negative green performance. Accordingly, the results of chapters 2, 3, 4, & 5 were revised to identify these cases. After revising the results of previous chapters, the cases adopted the ineffective strategy patterns were

Table 1. Cases Adopted Ineffective Green Strategy Patterns

Case	GHG1-Flight	GHG1-Ground	GHG2	Waste	Water
A6	Operating+				
NA3	Route+				
A3	Route+			Hybrid	
A4	Route+				
NA6	Design+	Design+			
NA4	Design+				
ME2	Design+				
ME1	Route+			Onboard & Ground	
A1	Design+				
NA1	Fuel				
NA5		Design+	Sustainable Energy+		Facilities & Buildings Water Saving
E3			Replacing Facilities	Onboard & Ground	

identified as presented in the Table 1.

Computing the Green Indexes of Ineffective Cases Before Improvement

Step 1: Deducing the percentage of change in green performance of all effective cases that adopted the same strategy pattern of ineffective case, from the percentage of change in the green performance of ineffective case.

Step 2: Computing the average difference in the performance for the ineffective case in comparison with the effective case.

Step 3: Dividing the average performance of ineffective case by the performance of most effective case under the same green indicator.

Step 4: Multiplying the relative performance of ineffective case by -5, which is the worst negative rating value of the negative rating scale. The result is the ineffective case green index.

Please see the example in the Appendix, Tables 14 and 15.

Recommending the Most Effective Strategy Pattern for the Ineffective Cases

Step 1: Identifying all effective strategy pattern alternatives.

Step 2: Identifying the most effective strategy alternative, which achieved the highest performance.

Step 3: identifying the impact of adopting the most effective strategy pattern on the performance of ineffective case as following:

a. Computing the change in the green performance of ineffective case.
b. Computing the green index again by dividing the performance of the case by the performance of best case, then, multiply the result by 5, which is the maximum effective rating scale.

Please see the example in the Appendix, Tables 16 and 17.

Identifying the Changes on the Green Strategy Pattern after Adopting the Most Effective Patterns

Step 1: Identifying the strategy pattern of the ineffective case before adopting the most effective pattern(s)

Step 2: identifying the changes on the pattern as a result of adopting the most effective pattern(s).

Step 3: identifying if the same pattern or a new pattern will be adopted.

Please see the example in the Appendix, Table 18.

RESEARCH RESULTS

This part of the chapter presents the main results of the empirical study. The aim of this part is to recommend the changes on the strategy patterns that adopted by the ineffective cases. This part started by presenting the green index before adopting the recommended patterns for each case, followed by the green index of each region. Then, the recommendation of the effective strategy patterns for each case was discussed. Next, the impact of these recommendations on the green indexes for each case and region was discussed, followed by discussing the impact of these recommendations on the general strategy patterns that could be adopted. In order to facilitate the reading of strategy patterns alternatives, Table 2 shows the acronyms of strategy patterns alternatives.

Airlines Green Index Before Improvement (Case Level Analysis)

Table 3 shows the green indexes for each airline. This table presenting the green index because of adopting the effective and ineffective green strategy patterns. It can be seen that the indexes values are positives and negatives. The positive values mean the case achieved improvement in the green performance. However, the negative values mean the case suffered from deterioration in green performance. The focus of this chapter is on the deteriorated indexes, since the concern is to shift the negative green performance to a better or a positive performance. It can be seen that the negative indexes are achieved by the positive and negative performed airlines. Three cases of the positive performed airlines suffered from deterioration in one green index (GHG1-Flight operations); these cases were NA1, A4, and NA3. However, all cases of the negative performed airlines suffered from deterioration in one or more green indicators. Nevertheless, the widely shared deteriorated negative indicator was GHG1-Flight operations. The average index of two indicators was negative, these indicators were GHG1-Flight operations and GHG1-Ground operations, and the index value was -0.17 and -0.16 respectively. The level of deterioration in the green indicators by the airlines range from low

Table 2. List of Acronyms Used in Presenting the Chapter Results

Acronym	Interpretation
A	Facilities and building accreditation
AS	Conformance with Airport fly quiet program
D	Aircraft Design
ES	Facilities and building energy saving
F	Flights fuel management
FAAS	Aircraft conformance with FAA noise standards
FO	Flights operating
FWR	Facilities and buildings water recycling and recovery
FWS	Facilities and buildings water saving
GHG1	Direct greenhouses gas emission
GHG2	Indirect greenhouse gas emission
HRU	Hazard wastes reusing
ICAOS	Aircraft conformance with ICAO Noise Standards
IRC	Industrial wastes recycling
IRU	Industrial wastes reusing
IUP	Industrial wastes upcycling
M	Flights maintenance management
MWR	Maintenance water recycling and recovery
MWS	Maintenance water saving
NM	Noise management
O	Flight Operating
ORC	Onboard and ground wastes recycling
ORP	Onboard and ground reduce using of paper
OUP	Onboard and ground wastes upcycling
R	Flights routes management
RF	Upgrading and replacing facilities
SE	Facilities and building sustainable energy
VD	Engine and vehicle design
VO	Engine and vehicle Operating
VM	Engine and vehicle maintenance
W	Flights weight management
WSM	Waste management
WTM	Water Management

to high, but the most frequent level was moderate, followed by low.

Table 3. Airline Green Index Before Improvement (Case Level Analysis)

Case	GHG1 Flight	GHG1 Ground	GHG2	Water	Waste	Average
Airlines Achieved Positive Index						
ME3	5.0 +H					5.0 +H
E7				5.0 +H		5.0 +H
E2					4.67 +H	4.7 +H
SA1	3.6 +H					3.6 +H
E5	3.5 +H					3.5 +H
NA2			2.35 +M			2.4 +M
NA1	-3.45 -H				5.0 +H	0.78 +L
A7	1.8 +M					1.8 +M
E1		2.6 +M	1.4 +L		0.75 +L	1.6 +L
E1	1.5 +L					1.5 +L
A4	-1.76 -M		0.94 +L	4.2 +H		1.1 +L
A5	0.7 +L				0.86 +L	0.8 +L
E4				0.82 +L	0.43 +L	0.6 +L
NA3	-0.09 -L	1.5 +L				0.7 +L
Airlines Achieved Negative Index						
A1	-2.06 -M		1.8 +M			-0.1 -L
A6	-0.53 -L					-0.5 -L
NA5	3.4 +M	-3.89 -H	-1.5 -L	-5.0 -H	2.95 +M	-0.8 –L
E3	3.3 +M		-5.0 -H		-3.31 -M	-1.67 -M
NA6	-3.26 -M	-2.5 -M	0.23 +L			-1.8 -M
ME1	-5.0 -H			1.2 +L	-3.33 -M	-2.4 -M
A3	-1.59 -L				-5.0 -H	-3.3 -M
NA4	-3.71 -H					-3.7 -H
ME2	-4.35 -H					-4.4 -H
Average of All Cases	-0.17 –L	-0.16 –L	0.65 +L	1.24 +L	0.34 +L	-0.17 –L

L: Low **M:** Moderate **H:** High
-: Deteriorated green performance
+: Improved green performance

Airlines Green Index Before Improvement (Regions Level Analysis)

Table 4 shows the airline green indexes for each region. This table presenting the green index because of adopting the effective and ineffective green strategy patterns. It can be seen that the indexes values are positives and negatives. The positive values mean the region achieved an improvement in the green performance. However, the negative means the region suffered from the deterioration in the green performance. The focus of this chapter is on the deteriorated indexes, since the concern is to shift the negative green performance to a better or a positive performance. It can be seen that the negative indexes are achieved by the positive and negative performed regions. Two regions of the positive performed suffered from deterioration in one or more green index, Europe suffered from deterioration in GHG2 indicator, however, Asia suffered from deterioration in GHG1-Flight operations and waste. However, all regions of the negative performed suffered from deterioration in more than on green indicator. Nevertheless, the widely shared deteriorated negative indicator was GHG1-Flight operations. The level of deterioration in the green indicators by the regions ranged from low to high, but the most frequent level was low.

Table 4. Airline Green Index Before Improvement (Regions Level Analysis)

Case	GHG1 Flight	GHG1 Ground	GHG2	Water	Waste	Average
Regions Achieved Positive Index						
Europe	2.77 +M	2.6 +M	-0.79 -L	2.91 +M	0.70 +L	1.65 +L
South America	3.6 +H					3.6 +H
Asia	-0.57 -L			4.2 +H	-2.07 -M	0.31 +L
Regions Achieved Negative Index						
North America	-1.20 -L	-1.20 -L	1.86 +L	-5.0 -H	3.75+H	-0.36 –L
Middle East	-1.45 -L			1.20 +L	-3.33 +M	-0.72 –L

L: Low **M:** Moderate **H:** High
-: Deteriorated green performance
+: Improved green performance

Recommended Airline Effective GHG1-Flight Operations Strategy

Table 5 shows the recommended effective strategy patterns to improve the GHG1-Flight operations emissions. It can be seen that these strategies are recommended for the following cases: NA1, A4, ME1, NA3, A3, NA6, NA4, ME2, A1 and A6. These cases adopted green operations strategies, but the adopted strategies were ineffective. Accordingly, the recommended strategies are modifications of the adopted strategies to have a better green performance.

Table 5. Recommended Airline Effective GHG1-Flight Operations Strategy

Case	Recommended Green Strategy		% Reduction in GHG1 Emissions by Each Action	Achieved % of Reduction in GHG Emission by Each case	Achieved Green Index by Each Case
	Recommended Strategy Pattern	Recommended Strategy Actions			
NA1	Fuel: F	Training in fuel saving	-25%	**NA1**: 0.15%	**NA1**: -0.02 -L
A4 ME1 NA3 A3	Rout+: R-O-M	Broadcasting technology for communication everywhere Monitoring actual flight path and relevant data Reduced use of auxiliary power unit Altitude optimization Reducing landing time Engine Washing	-5.4% -6.3% -2.4% -2.8% -3.3% -6.9%	**A4:** -14% **ME1:** +9.0% **NA3:** -16% **A3:** -15%	**A4:** 1.94 +M **ME1:** -1.25 -L **NA3:** 2.16 +M **A3:** 2.11 +M
NA6 NA4 ME2 A1	Design+: D-R-O-W-F-M	Shark let wing tips Split scimitar winglets PC-calculated flight path Broadcasting technology for communication everywhere Route optimization Monitoring actual flight path and relevant data Acceleration altitude Accurate planning of flight Optimal water potable carried Using EFB (Electronic Flight Bag) Using onboard tablets Light meal trolleys Eliminating magazines Light weight cargo containers Restricting fuel overload Engine washing	-2.4% -2.8% -2.4% -2.4% -3% -2.7% -2.5% -2.5% -2% -2.5% -2.5% -2.3% -2.4% -2.5% -3% -2.3%	**NA6:** -14% **NA4:** -11% **ME2:** -6% **A1:** -23%	**NA6:** 1.92 +M **NA4:** 1.47 +L **ME2:** 0.82 +L **A1:** 3.13 +M
A6	Operating+: O	Reduced use of auxiliary power unit Landing continuous descent operations	-2.7% -2.7%	**A6:** -1.5%	**A6:** 0.20 +L

L: Low **M:** Moderate **H:** High

Moreover, the strategy pattern title that recommended is the same as the adopted, but the recommended one is the most effective pattern alternatives. The reason behind recommending alternative of the same pattern sine the reporting of the most effective pattern was according to the comparison with these ineffective cases. Accordingly, the ineffective cases are belonging to the same cluster or group of the most effective cases.

The recommended strategy patterns are: fuel (F) pattern, which is recommended for NA1. Route+ (R-O-M) pattern, which is recommended for A4, ME1, NA3 and A3. Design+ (D-R-O-W-F-M) pattern, which is recommended for NA6, NA4, ME2 and A1. Operating+ (O) pattern, which is recommended for A6. The list of recommended strategy actions is corresponding each strategy pattern. The impact of these actions ranges from -2% to -25%. The impact of adopting these strategy patterns on reducing GHG1- Flight operations emissions differs. The range of reduction is between -1.5% to -23%. In addition, the adopting of the recommended strategy pattern will improve green index of each case. The most of cases will achieve a moderate index value, these cases are: A4, NA3, A3, NA6, A1 and A6. Two cases will not achieve a positive index, but the negative performance will be minimized, these cases are: NA1 and ME1, the achieved negative index will be low by a value of -0.02 and -1.25 respectively.

Recommended Airline Effective GHG1-Ground Operations Strategy

Table 6 shows the recommended effective strategy pattern to improve the GHG1-Ground operations emissions. It can be seen that this strategy is recommended for the cases, NA6 and NA5. These cases adopted green operations strategy, but the adopted strategy was ineffective. Accordingly, the recommended strategy is a modification of the adopted strategy to have a better green performance. Moreover, the strategy pattern title that recommended is the same as adopted, but the recommended is the most effective pattern alternative. The reason behind recommending alternative of the same pattern sine the reporting of most effective pattern was according to the comparison with these ineffective cases. Accordingly, the ineffective cases are belonging to the same cluster or group of the most effective cases.

The recommended strategy pattern is: design+ (VD-VM) pattern, which is recommended for both cases. The list of recommended strategy actions is corresponding the strategy pattern. The impact of these actions is the same,

Table 6. Recommended Airline Effective GHG1-Ground Operations Strategy

Case	Recommended Green Strategy		% Reduction in GHG1 Emissions by Each Action	Achieved % of Reduction in GHG Emission by Each case	Achieved Green Index by Each Case
	Recommended Strategy Pattern	Recommended Strategy Actions			
NA6 **NA5**	Design+: VD-VM	Using electrical carts or tractors (trucks). Washing engine with less water. Cleaning vehicles and equipment.	-6.6% -6.6% -6.6%	**NA6**: -1.50% **NA5**: +7.92%	**NA6**: 0.21 +L **NA5**: -1.1 -L

L: Low **M**: Moderate H: High

which is -6.6%. The impact of adopting this strategy pattern on reducing GHG1-Ground operations is -1.50% if it will be adopted by the case NA6. However, the adopting of this strategy will not impact the improvement of performance of NA5, but the achieved GHG1-Ground operations was positive by 7.4%. In addition, the green index of NA6 will be improved to reach the low level by 0.21. However, the achieved index by the case NA6 will negative, the index value will be -1.1.

Recommended Airline Effective GHG2 Operations Strategy

Table 7 shows the recommended effective strategy patterns to improve the GHG2 emissions. It can be seen that these strategies are recommended for the cases, NA5 and E3. These cases adopted green operations strategies, but the adopted strategies were ineffective. Accordingly, the recommended strategies are modifications of the adopted strategies to have a better green performance. Moreover, the strategy patterns titles recommended are the same as adopted, but the recommended strategies are the most effective pattern alternatives. The reason behind recommending alternatives of the same pattern sine the reporting of most effective pattern was according to the comparison with these ineffective cases. Accordingly, the ineffective cases are belonging to the same cluster or group of the most effective cases.

The recommended strategy patterns are: sustainable energy+ (SE-RF) pattern, which is recommended for the case NA5, and replacing facilities (RF) pattern which was recommended for E3. The list of recommended strategy actions are corresponding the strategy patterns. The impact of these actions on reducing GHG2 emissions is -3.25% for each action of sustainable energy+. However, the impact of the only action of replacing facilities pattern

Table 7. Recommended Airline Effective GHG2 Operations Strategy

Case	Recommended Green Strategy		% Reduction in GHG2 Emissions by Each Action	Achieved % of Reduction in GHG2 Emission by Each case	Achieved Green Index by Each Case
	Recommended Strategy Pattern	Recommended Strategy Actions			
NA5	Sustainable Energy+: SE-RF	Using solar panel. Upgrading the copiers, fax and printer to energy-friendly ones. Upgrading and adjusting building control and automation system to conserve electricity. Replacing power equipment at several radar sites.	-3.25% -3.25% -3.25% -3.25%	**NA5**: +0.28%	**NA5**: -0.03 -L
E3	Replacing Facilities: RF	Upgrading the copiers, fax and printer to energy-friendly ones	-43%	**E3**: +0.4%	**E3**: -0.05 -L

L: Low **M**: Moderate **H**: High

is -43%. The adopting of these strategies will improve the GHG2 emissions, but the emission values are still positive. The achieved GHG2 emissions by the case NA5 will be +0.28%. However, the achieved emissions by E3 will be +0.41%. In addition, the green index of NA5 and E3 will be negative to reach the value of -0.03 and -0.05 respectively.

Recommended Airline Effective Wastes Management Strategy

Table 8 shows the recommended effective strategy patterns to reduce wastes. It can be seen that these strategies are recommended for the following cases: ME1, E3 and A3. These cases adopted green operations strategies, but the adopted strategies were ineffective. Accordingly, the recommended strategies are modifications of the adopted strategies to have a better green performance. Moreover, the strategy pattern title recommended is the same as adopted, but the recommended is the most effective pattern alternatives. The reason behind recommending alternative of the same pattern sine the reporting of most effective pattern was according to the comparison with these ineffective cases. Accordingly, the ineffective cases are belonging to the same cluster or group of the most effective cases.

The recommended strategy pattern is: onboard and ground wastes management, which is recommended for all cases. The list of recommended strategy actions is corresponding the strategy pattern. The impact of these actions ranges from -7.5% to -17.79%. The impact of adopting this strategy

Table 8. Recommended Airline Effective Wastes Management Strategy

Case	Recommended Green Strategy		% Reduction in Wastes by Each Action	Achieved % of Reduction in Wastes by Each case	Achieved Green Index by Each Case
	Strategy Pattern	Strategy Actions			
ME1 **E3** **A3**	Onboard and ground: ORC-ORP	Auditing internal recycling. Recycling paper. Recycling plastic. Recycling cans. Using electronic airway bills for freight documents. Using electronic boarding passes. Increasing the number of recycling stations. Adopting a recycling tracking system. Recycling oils from restaurant. Burning it to recover energy.	-17.79% -7.50% -7.50% -7.50% -14.41% -14.41% -15.67% -16.75% -15.67% -15.67%	**ME1**: -44% **E3**: -44% **A3**: -27%	**ME1**: 3.20 +M **E3**: 3.20 +M **A3**: 1.95 +M

L: Low **M**: Moderate **H**: High

pattern on reducing the wastes differs. The range of reducing in wastes is between -27% to -44%. In addition, the adopting of the recommended strategy pattern will improve the green index of each case. All cases will achieve a moderate index value. The achieved indexes by these cases will be moderate by 3.20 for ME1 and E3. However, the achieved index by A3 will be 1.95.

Recommended Airline Effective Water Management Strategy

Table 9 shows the recommended effective strategy pattern to reduce the water consumption. It can be seen that this strategy is recommended for the case NA5. This case adopted a green operations strategy, but the adopted strategy was ineffective. Accordingly, the recommended strategy is modifications of the adopted strategy to have a better green performance. Moreover, the strategy pattern title recommended is the same as adopted, but the recommended is the most effective pattern alternative. The reason behind recommending alternative of the same pattern sine the reporting of most effective pattern was according to the comparison with these ineffective cases. Accordingly, the ineffective cases are belonging to the same cluster or group of the most effective cases.

The recommended strategy pattern is: facilities and buildings water recovery and recycling. The list of the recommended strategy actions is corresponding the strategy pattern. The impact of these actions is the same,

Table 9. Recommended Airline Effective Water Management Strategy

Case	Recommended Green Strategy		% Reduction in Water by Each Action	Achieved % of Reduction in Water by Each case	Achieved Green Index by Each Case
	Strategy Pattern	Strategy Actions			
NA5	Facilities and Buildings Water Recovery and Recycling (FWR)	Using a rain water recovery system Recycling water used in the cooling towers of the building	-11% -11%	**NA5**: -13%	**NA5**: 2.94 +M

L: Low **M**: Moderate **H**: High

which is -11% for each action. The impact of adopting this strategy pattern on reducing the wastes by the case NA5 will be -13%. In addition, the adopting of recommended strategy pattern will improve the green index of NA5 to reach a moderate value by 2.94.

Airlines Green Index After Improvement (Case Level Analysis)

Table 10 shows the impact of adopting the recommended airline effective green strategies (improvement) on the green indexes of each case, and the average changes in the green indexes. It can be seen that the green indexes for all cases after improvement will be positive, so no more deteriorated cases in green performance after making improvement. About half of the improved cases will achieve a moderate green index, these cases are: NA1, A4, NA3, A1, E3 and A3. However, the remaining improved cases will achieve a low green index, these cases are: A6, NA5, NA6, ME1, NA4 and ME2. The range of change in green index will be between +0.7 to +5.33. On the other hand, the average overall index of all green indicators will be improved to be positive. The achieved index will be low for three green indicators, which are GHG1-Flight, GHG1-Ground, and GHG2. However, a moderate index will be achieved for two indicators, which are waste and water.

Airlines Green Index After Improvement (Regional Level Analysis)

Table 11 shows the green indexes will be achieved by the regions as a result of adopting the recommended effective green strategy (improvement). It can be seen that all indexes of all regions will be improved to be positive. The

Table 10. Airline Green Index After Improvement (Case Level Analysis)

Case	GHG1 Flight	GHG1 Ground	GHG2	Water	Waste	Index Before Improvement	Index After Improvement	Change in Index
ME3	5 +H					5.0 +H	5 +H	
E7				5 +H		5.0 +H	5 +H	
E2					4.67 +H	4.7 +H	4.67 +H	
SA1	3.6 +M					3.6 +M	3.6 +M	
E5	3.5 +M					3.5 +M	3.5 +M	
NA2			2.35 +M			2.35 +M	2.35 +M	
NA1	-0.02 -L				5 +H	0.78 +L	2.39 +M	+1.13
A7	1.8 +M					1.8 +M	1.8 +M	
E1		2.6 +M	1.4 +L		0.75 +L	1.6 +L	1.58 +L	
E1	1.5 +L					1.5 +L	1.5 +L	
A4	1.94 +M		0.94 +L	4.2 +H		1.1 +L	2.36 +M	+1.26
A5	0.7 +L				0.86 +L	0.8 +L	0.78 +L	
E4				0.82 +L	0.43 +L	0.6 +L	0.63 +L	
NA3	2.16 +M	1.5 +L				0.7 +L	1.83 +M	+1.13
A1	3.13 +M		1.8 +M			-0.1 -L	2.47 +M	+2.57
A6	0.2 +L					-0.5 -L	0.2 +L	+0.70
NA5	3.4 +M	-1.1 –L	-0.03 -L	2.94 +M	2.95 +M	-0.8 –L	1.63 +L	+2.43
E3	3.3 +M		-0.05 -L		3.2 +M	-1.67 -L	2.15 +M	+3.82
NA6	1.92 +L	0.21 +L	0.23 +L			-1.8 –M	0.79 +L	+2.59
ME1	-1.25 -L			1.2 +L	3.2 +M	-2.4 –M	1.05 +L	+3.45
A3	2.11 +M				1.95+L	-3.3 -M	2.03 +M	+5.33
NA4	1.47 +L					-3.7 -H	1.47 +L	+5.17
ME2	0.82 +L					-4.4 -H	0.82 +L	+5.22
Average Before Improvement	-0.17 -L	-0.16 –L	0.65 +L	1.24 +L	0.34 +L	-0.17 –L		
Average After Improvement	1.96 +L	0.94 +L	1.46 +L	2.83 +M	2.56 +M	0.72 +L		

L: Low **M:** Moderate **H:** High
-: Deteriorated green performance
+: Improved green performance

average index of Europe and Asia will be shifted from low to moderate. The average index of North America and the Middle East will be shifted from a low negative level to a moderate positive level.

Table 11. Airline Green Index After Improvements (Regional Level Analysis)

Region	Achieved Index	GHG1 Flight	GHG1 Ground	GHG2	Water	Waste	Average
Europe	Before Improvement	2.77 +M	2.06 +M	-0.79 -L	2.91 +M	0.70 +L	1.65 +L
	After Improvement	2.77 +M	2.06 +M	0.68 +L	2.91 +M	2.26 +M	2.13 +M
South America	Before Improvement	3.60 +H					3.60 +H
	After Improvement	3.60					3.60 +H
Asia	Before Improvement	-0.57 -L			4.20 +H	-2.07 -M	0.31 +L
	After Improvement	1.65 +L		1.37 +L	4.20 +H	1.41 +L	2.16+M
North America	Before Improvement	-1.20 -L	-1.20 -L	1.86 +M	-5.0 -H	3.75+H	-0.36 –L
	After Improvement	1.79 +M	0.20 +L	1.89 +M	2.94 +M	3.98 +H	2.16 +M
Middle East	Before Improvement	-1.45 –L			1.20 +L	-3.33 -M	-0.72 –L
	After Improvement	1.52 +L			1.20 +L	3.20 +M	1.97+M

L: Low **M:** Moderate **H:** High
-: Deteriorated green performance
+: Improved green performance

Changes in Effective Green Operations Strategy Patterns

Table 12 shows the changes in the effective green operations strategy patterns as a result of adopting the most effective strategy alternatives under each green indicator. It can be seen that the strategy patterns of six cases will be changes, the pattern of NA1 will be changed from waste management+ to GHG, waste & noise. The pattern of ME1 will be changed from water management to GHG &water management+. The pattern of NA5 will be changed from GHG, waste & noise to do all pattern. However, the noise management pattern of three cases A3, A6 and ME2 will be changed to GHG, waste & noise, and will be changed to GHG & Noise Management for cases A6 and ME2. The remaining cases will adopt the same pattern but there will be changes in the strategy alternatives under each indicator.

Table 12. Changes in Effective Airline Green Operations Strategy Patterns

Case	Strategy pattern before improvement	Strategy pattern after improvement	Pattern Alternative	Index of new pattern
NA1	Waste Management+	GHG, Waste & Noise	**GHG1:**F --- **WSM:** ORC-ORP	2.39 +M
ME1	Water Management	GHG & Water Management+	**GHG1:** R-O-M--- -**WTM**: FWS--- **WSM:** ORC-ORP	1.05 +L
E3	GHG &Noise Management	GHG &Water Management+	**GHG1:** F---**GHG2:**RF--- **WTM:** ORC-ORP---**NM:** D-ICAO-FO	2.15 +M
NA3	GHG &Noise Management	GHG &Noise Management	**GHG1:** R-O-M-VD-VO--- **GHG2:** A---**NM:**FAAS	1.83 +M
NA6	GHG &Noise Management	GHG &Noise Management	**GHG1**: D-R-O-W-F-M--- **GHG2:** A-RF---**NM:**D	0.78 +L
A1	GHG Emissions	GHG Emissions	**GHG1:**D-R-O-W-F-M--- **GHG2:** ES-A-SE	2.17 +M
NA5	GHG, Waste & Noise	Do All	**GHG1:** R-F- VD-VR --- **GHG2:**SE-RF **WSM:** ORC-OUP-IRC-IUP-IRU-HRU---**WTM:** FWR---**NM:**D-ICAOS	1.63 +L
A4	GHG & Water Management+	GHG & Water Management+	**GHG1:** R-O-M---**GHG2:** ES---**WTM:**MWR-FWR-FWS	2.36 +M
A3	Noise Management	GHG, Waste & Noise	**GHG1:** R-O-M--- **WSM:**ORC-ORP--- **NM:**ICOS	2.03 +M
A6	Noise Management	GHG &Noise Management	**GHG1**:O---**NM:**FO	0.20 +L
ME2	Noise Management	GHG &Noise Management	**GHG1:**D-R-O-W-F-M--- **NM:**ICAOS	0.82 +L

L: Low **M:** Moderate **H:** High

Modified Airline Effective Green Operations Strategy Patterns after Improvement

Table 13 shows the modified airline effective green operations strategy patterns after improvement. In order to trace the changes please see the original effective green operations strategy patterns before improvement, which is appended in the Appendix, Table 19. The number of original patterns before improvement was eight. However, the number of effective patterns after improvement will be seven. The recommended airline green operation strategies are: waste management+, GHG & noise management, GHG emissions, GHG, waste

Table 13. Modified Airline Effective Green Operations Strategy Patterns

Pattern	Case ID	Pattern Alternatives	Performance
Waste Management+	ME1 E2 E4	**WSM:** ORC-ORP **WSM:** ORC-HRU---**NM:** D-ICAOS-AS-FO WSM:ORC---**WTM:**FWS	4.64 (H) 4.76 (H) 0.46 (L) **2.56 (M)**
GHG &Noise Management	ME3 SA1 NA3 NA6 A6 ME2	**GHG1:** D-R-O-W-F-M---**NM:**AD **GHG1:** R-O-M---**NM:**D-ICAOS **GHG1:** R-O-M-VD-VO---**GHG2:** A---**NM:**FAAS **GHG1:** D-R-O-W-F-M---**GHG2:** A-RF---**NM:**D **GHG1:**O---**NM:**FO **GHG1:**D-R-O-W-F-M---**NM:**ICAOS	5.00 (H) 3.63 (H) 1.83 (M) 0.78 (L) 0.20 (L) 0.82 (L) **2.54 (M)**
GHG Emissions	E5 A1	**GHG1:** R-O-W---**GHG2:** SE-RF **GHG1:**D-R-O-W-F-M---**GHG2:** ES-A-SE	2.21 (H) 2.17 (M) **2.19 (M)**
GHG, Waste & Noise	A5 NA2 A3 NA1	**GHG1:**O---**WSM:**ORC---**NM:**D **GHG2:** ES-A---**WSM:** ORC-ORP- IUP-IRU---**NM:** ICAOS **GHG1:** R-O-M---**WSM:**ORC-ORP---**NM:**ICOS **GHG1:**F--- **WSM:**ORC-ORP	0.79 (L) 2.83 (M) 2.03 (M) 2.39 (M) **2.26 (M)**
GHG & Water Management+	A4 A7 E3 ME1	**GHG1:** R-O-M---**GHG2:** ES---**WTM:**MWR-FWR-FWS **GHG1:**D-R-M- ---**WTM:**FWR---**NM:**ICAOS **GHG1:** F---**GHG2:**RF---**WTM:** ORC-ORP---**NM:** D-ICAO-FO **GHG1:** R-O-M--- **WTM:** FWS--- **WSM:** ORC-ORP	2.36 (M) 3.38 (H) 2.15 (M) 1.05 (L) **3.03 (M)**
Do All	E1 NA5	**GHG1:** O-F- VD-VM ---**GHG2:**ES-SE-RF---**WSM:** OUP---**WTM:** MWR-MWS-FWS---**NM:** D-ICAOS-FO **GHG1:** R-F- VD-VR --- **GHG2:**SE-RF **WSM:** ORC-OUP-IRC-IUP-IRU-HRU---**WTM:** FWR---**NM:**D-ICAOS	0.90 (L) 1.63 (L) **1.27 (L)**
Noise Management	A2	**NM:** D-FO	----

(L): Low **(M):** Moderate **(H):** High
Note: to have a better insight about the strategy patterns' acronyms, please see Table 2.

& noise, GHG & water management +, do all, and noise management. The eliminated pattern will be water management. The most frequent strategy pattern is GHG & noise management, which is the same highly frequent pattern before improvement. Also the green index achieved by the strategy patterns before improvement was low for two patterns and moderate for the remaining. However, after improvement the green indexes will be moderate for six patterns and low for one pattern.

CONCLUSION

The aim of this chapter was to recommend changes in the ineffective green strategy patterns to have more effective green strategy patterns. The previous studies have not raised this research issue. The main results of this study were: the green indexes for all cases after improvement will be positive, so no more deteriorated cases in green performance after making improvement. About half of the improved cases will achieve a moderate green index. The average overall index of all green indicators will be improved to be positive. All indexes of all regions will be improved to be positive. The average index of Europe and Asia will be shifted from low to moderate. The average index value of North America and the Middle East will be shifted from the low negative level to the moderate positive. The number of original patterns before improvement was eight. However, the number of effective patterns after improvement will be seven. Also the green index achieved by the strategy patterns before improvement was low for two patterns and moderate for the remaining. However, after improvement the green indexes of all effective strategy patterns will be moderate for six patterns, and low for one pattern. The results of this chapter will benefit the academics and the industry experts alike. The academic have in-depth results about the most effective strategy patterns that could be used to develop more extended theoretical models. The experts in the industry have a clear idea about the recommended effective strategy patterns that could be adopted.

REFERENCES

Migdadi, Y. K. A. A. (2016). Identifying the best practices in green operations strategy of leading mobile phone producers. *International Journal of Business Excellence*, *9*(1), 92–112. doi:10.1504/IJBEX.2016.073377

Migdadi, Y. K. A. A. (2018). Identifying the best practices of airlines' green operations strategy: A cross-regional worldwide survey. *Environmental Quality Management*, *28*(1), 21–32. doi:10.1002/tqem.21575

Migdadi, Y. K. A. A. (2019). Identifying the effective taxonomies of airline green operations strategy. *Management of Environmental Quality*.

Migdadi, Y. K. A. A. (2020). Green Operational Strategy for Airlines: Content and Regional Analysis. In Cases on Green Energy and Sustainable Development (pp. 193-229). IGI Global.

Migdadi, Y. K. A. A., & Elzzqaibeh, D. A. S. I. (2018). The evaluation of green manufacturing strategies adopted by ISO 14001 certificate holders in Jordan. *International Journal of Productivity and Quality Management, 23*(1), 90–109. doi:10.1504/IJPQM.2018.088610

Migdadi, Y. K. A. A., & Omari, A. A. (2019). Identifying the best practices in green operations strategy of hospitals. *Benchmarking: An International Journal, 26*(4), 1106–1131. doi:10.1108/BIJ-09-2017-0242

APPENDIX

How to Compute the Green Index of Ineffective Cases Before Improvement

Table 14. Example on How to Compute the Green Index of Ineffective Cases before Improvement: Step1

Ineffective Case			Effective Cases			Step1: Deducting the percentage of change in green performance of effective from ineffective cases
Case	Strategy Pattern	Percentage of Change in GHG1 Emissions	Case	Strategy Pattern	Percentage of Change in GHG1 Emissions	
A6	Operating+	0.3%	A5	Operating+	-5%	0.3%- (-5%)= +5.3%
			E1	Operating+	-2%	0.3%- (-2%)= +2.3%

Table 15. Example on How to Compute the Green Index of Ineffective Cases before Improvement: Step 2, 3, 4.

Procedure	Calculations
Step 2: Computing the average differencing the performance of ineffective case	(5.3%+2.3%)/2= 3.75%
Step 3: dividing the average performance of ineffective case by the performance of the most effective case adopted the same green indicator	(3.75% / 36%)= 0.10
Step 4: multiplying the relative performance of ineffective case by -5. The result is the ineffective case green index	0.10 X -5= -0.50 L

How to Identify the Most Effective Strategy Pattern

Table 16. Example on How Identify the Most Effective Strategy Pattern

<table>
<tr><th colspan="4">Step 1: Identifying all effective strategy pattern alternatives</th><th rowspan="2">Step 2: identifying the most effective alternative</th></tr>
<tr><th>Case</th><th>Pattern Alternative</th><th>Significant Action Taken</th><th>Actions Impact</th></tr>
<tr><td>A5</td><td>Operating (O)</td><td>Reduced use of auxiliary power unit
Landing continuous descent operations</td><td>-2.7%
-2.7%</td><td rowspan="2">This is the most effective pattern since its total impact -5.4% is more than the impact of the second pattern -2.4%</td></tr>
<tr><td colspan="3">Total Pattern Impact</td><td>-5.4%</td></tr>
<tr><td>E1</td><td>Operating (O)
Fuel (F)</td><td>Reduced use of auxiliary power unit
Reducing aircraft take off time
Reducing flight launching time
Reduced taxiing time
Using biofuels</td><td>-0.5%
-0.5%
-0.5%
-0.5%
-0.4%</td><td rowspan="2"></td></tr>
<tr><td colspan="3">Total Pattern Impact</td><td>-2.4%</td></tr>
</table>

Table 17. Example on Step 3: How to Identify the Impact of Adopting the Most Effective Strategy Pattern on the Performance of Ineffective Case

Case	Performance before adopting the best pattern	Step a: Performance after adopting the best pattern	Step b: Computing the green index after adopting the best pattern
A6	+3.75%	+3.75% - 5.4% = -1.55%	(-1.55%/-36%)*5= 0.22 +L

How to Identify the Changes in Effective Green Operations Strategy Pattern

Table 18. Example How to Identify the Changes in Effective Green Operations Strategy Pattern After adopting the Most Effective Patterns

Case	Step 1: Before Adopting the Effective Pattern		Step 1: After Adopting the Effective Pattern	
	Strategy Pattern	**Alternative**	**Strategy Pattern**	**Alternative**
A6	Noise Management	**NM:**ICAOS	**Step 3:** new pattern GHG & Noise Management	**Step 2:** changes as shaded **GHG1:**D-R-O-W-F-M---**NM:**ICAOS

The Effective Green Operations Strategy Patterns Before Improvement

Table 19. Airline Effective Green Operations Strategy Patterns Before Improvement (The Results of Chapter 7)

Pattern	Case ID	Pattern Alternatives	Performance
Water Management	ME1	**WTM:** FWS	1.20 (L) **1.20 (L)**
Waste Management+	E2 E4 NA1	**WSM:** ORC-HRU---**NM:** D-ICAOS-AS-FO **WSM:**ORC---**WTM:**FWS **WSM:**ORC-ORP	4.67 (H) 0.46 (L) 5.00 (H) **3.41 (M)**
GHG &Noise Management	E3 ME3 SA1 NA3 NA6	**GHG1:** F---**NM:** D-ICAO-FO **GHG1:** D-R-O-W-F-M---**NM:**AD **GHG1:** R-O-M---**NM:**D-ICAOS **GHG1:** VD-VO---**GHG2:** A---**NM:**FAAS **GHG2:** A-RF---**NM:**D	3.32(M) 5.00 (H) 3.63 (H) 1.44 (L) 0.23 (L) **2.73 (M)**
GHG Emissions	E5 A1	**GHG1:** R-O-W---**GHG2:** SE-RF **GHG2:** ES-A-SE	2.21 (M) 1.80 (M) **2.00 (M)**
GHG, Waste & Noise	A5 NA2 NA5	**GHG1:**O---**WSM:**ORC---**NM:**D **GHG2:** ES-A---**WSM:** ORC-ORP- IUP-IRU---**NM:** ICAOS **GHG1:** R-F---**WSM:** ORC-OUP-IRC-IUP-IRU-HRU---**NM:**D-ICAOS	0.79 (L) 2.83 (M) 3.17 (M) **2.26 (M)**
GHG & Water Management+	A4 A7	**GHG2:** ES---**WTM:**MWR-FWR-FWS **GHG1:**D-R-M---**WTM:**FWR---**NM:**ICAOS	2.67 (M) 3.38 (H) **3.03 (M)**
Do All	E1	**GHG1:** O-F- VD-VM ---**GHG2:**ES-SE-RF---**WSM:** OUP---**WTM:** MWR-MWS-FWS---**NM:** D-ICAOS-FO	**0.90 (L)**
Noise Management	A2 A3 A6 ME2	**NM:** D-FO **NM:**ICOS **NM:**FO **NM:**ICAOS	---- ---- ---- ----

(L): Low **(M):** Moderate **(H):** High
Note: to have a better insight about the strategy patterns' acronyms, please see Table 2.

Related Readings

To continue IGI Global's long-standing tradition of advancing innovation through emerging research, please find below a compiled list of recommended IGI Global book chapters and journal articles in the areas of waste management, environmental sustainability, and air pollution. These related readings will provide additional information and guidance to further enrich your knowledge and assist you with your own research.

Affam, A. C., & Chaudhuri, M. (2019). Comparative Study of Advance Oxidation Processes for Treatment of Pesticide Wastewater. In H. Aziz & S. Abu Amr (Eds.), *Advanced Oxidation Processes (AOPs) in Water and Wastewater Treatment* (pp. 261–323). Hershey, PA: IGI Global. doi:10.4018/978-1-5225-5766-1.ch011

Ahmedi, F., Ahmedi, L., O'Flynn, B., Kurti, A., Tahirsylaj, S., Bytyçi, E., ... Salihu, A. (2018). InWaterSense: An Intelligent Wireless Sensor Network for Monitoring Surface Water Quality to a River in Kosovo. *International Journal of Agricultural and Environmental Information Systems*, *9*(1), 39–61. doi:10.4018/IJAEIS.2018010103

Akash, P. V. M., Babu, N., & Navneet, (2018). Bioremediation of Environmental Pollutants. In V. Pathak, & Navneet (Eds.), Handbook of Research on Microbial Tools for Environmental Waste Management (pp. 80-104). Hershey, PA: IGI Global. doi:10.4018/978-1-5225-3540-9.ch005

Akbar, N. A., Matsin, Z. B., & Ramli, S. F. (2019). Groundwater Treatment via Ozonation. In H. Aziz & S. Abu Amr (Eds.), *Advanced Oxidation Processes (AOPs) in Water and Wastewater Treatment* (pp. 200–225). Hershey, PA: IGI Global. doi:10.4018/978-1-5225-5766-1.ch009

Ali, D. A., Palaniandy, P., & Feroz, S. (2019). Advanced Oxidation Processes (AOPs) to Treat the Petroleum Wastewater. In H. Aziz & S. Abu Amr (Eds.), *Advanced Oxidation Processes (AOPs) in Water and Wastewater Treatment* (pp. 99–122). Hershey, PA: IGI Global. doi:10.4018/978-1-5225-5766-1.ch005

Amorim, C. L., Moreira, I. S., Duque, A. F., van Loosdrecht, M. C., & Castro, P. M. (2017). Aerobic Granular Sludge: Treatment of Wastewaters Containing Toxic Compounds. In Á. Val del Río, J. Campos Gómez, & A. Mosquera Corral (Eds.), *Technologies for the Treatment and Recovery of Nutrients from Industrial Wastewater* (pp. 231–263). Hershey, PA: IGI Global. doi:10.4018/978-1-5225-1037-6.ch009

Anderson, S., & Rahman, P. K. (2018). Bioprocessing Requirements for Bioethanol: Sugarcane vs. Sugarcane Bagasse. In V. Pathak, & Navneet (Eds.), Handbook of Research on Microbial Tools for Environmental Waste Management(pp. 48-56). Hershey, PA: IGI Global. doi:10.4018/978-1-5225-3540-9.ch003

Ara, T., Bashir, R., Chisti, H., & Rangreez, T. A. (2019). Wastewater Pollution From the Industries. In A. Hussain & S. Ahmed (Eds.), *Advanced Treatment Techniques for Industrial Wastewater* (pp. 98–113). Hershey, PA: IGI Global. doi:10.4018/978-1-5225-5754-8.ch007

Aris, A., Jusoh, M. N., & Wahab, N. S. (2019). Applications of Advanced Oxidation Processes in Palm Oil Mill Effluent Treatment. In H. Aziz & S. Abu Amr (Eds.), *Advanced Oxidation Processes (AOPs) in Water and Wastewater Treatment* (pp. 123–149). Hershey, PA: IGI Global. doi:10.4018/978-1-5225-5766-1.ch006

Aziz, N. A., & Palaniandy, P. (2019). Photocatalysis (TiO2/Solar) in Water and Wastewater Treatment. In H. Aziz & S. Abu Amr (Eds.), *Advanced Oxidation Processes (AOPs) in Water and Wastewater Treatment* (pp. 171–199). Hershey, PA: IGI Global. doi:10.4018/978-1-5225-5766-1.ch008

Babu, N., & Pathak, V. M. Akash, & Navneet, (2018). Biosorption of Heavy Metals: Biological Approach to Control the Industrial Waste. In V. Pathak, & Navneet (Eds.), Handbook of Research on Microbial Tools for Environmental Waste Management (pp. 270-281). Hershey, PA: IGI Global. doi:10.4018/978-1-5225-3540-9.ch013

Bagheri, S., & Julkapli, N. M. (2017). Biomass-Derived Activated Carbon: Synthesis, Functionalized, and Photocatalysis Application. In T. Saleh (Ed.), *Advanced Nanomaterials for Water Engineering, Treatment, and Hydraulics* (pp. 162–199). Hershey, PA: IGI Global. doi:10.4018/978-1-5225-2136-5.ch007

Banerjee, P., Mukhopadhayay, A., & Das, P. (2016). Advances in Bioremediation for Removal of Toxic Dye from Different Streams of Wastewater. In A. Rathoure & V. Dhatwalia (Eds.), *Toxicity and Waste Management Using Bioremediation* (pp. 266–278). Hershey, PA: IGI Global. doi:10.4018/978-1-4666-9734-8.ch013

Basak, B., & Dey, A. (2016). Bioremediation Approaches for Recalcitrant Pollutants: Potentiality, Successes and Limitation. In A. Rathoure & V. Dhatwalia (Eds.), *Toxicity and Waste Management Using Bioremediation* (pp. 178–197). Hershey, PA: IGI Global. doi:10.4018/978-1-4666-9734-8.ch009

Basheer, T., & Umesh, M. (2018). Valorization of Tannery Solid Waste Materials Using Microbial Techniques: Microbes in Tannery Solid Waste Management. In V. Pathak, & Navneet (Eds.), Handbook of Research on Microbial Tools for Environmental Waste Management (pp. 127-145). Hershey, PA: IGI Global. doi:10.4018/978-1-5225-3540-9.ch007

Becker, J. A., & Stefanakis, A. I. (2016). Pharmaceuticals and Personal Care Products as Emerging Water Contaminants. In A. McKeown & G. Bugyi (Eds.), *Impact of Water Pollution on Human Health and Environmental Sustainability* (pp. 81–100). Hershey, PA: IGI Global. doi:10.4018/978-1-4666-9559-7.ch004

Belmonte, M., Fajardo, C., Toledo-Alarcón, J. B., Heredia, D. V., Jorquera, L., Méndez, R., ... Ruiz-Filippi, G. (2017). Autotrophic Denitrification Processes. In Á. Val del Río, J. Campos Gómez, & A. Mosquera Corral (Eds.), *Technologies for the Treatment and Recovery of Nutrients from Industrial Wastewater* (pp. 147–173). Hershey, PA: IGI Global. doi:10.4018/978-1-5225-1037-6.ch006

Bender, S. R. (2016). Floodplain Infrastructure and the Toxic Tide. In A. McKeown & G. Bugyi (Eds.), *Impact of Water Pollution on Human Health and Environmental Sustainability* (pp. 150–173). Hershey, PA: IGI Global. doi:10.4018/978-1-4666-9559-7.ch007

Benjamin, S. R., de Lima, F., & Rathoure, A. K. (2016). Genetically Engineered Microorganisms for Bioremediation Processes: GEMs for Bioremediaton. In A. Rathoure & V. Dhatwalia (Eds.), *Toxicity and Waste Management Using Bioremediation* (pp. 113–140). Hershey, PA: IGI Global. doi:10.4018/978-1-4666-9734-8.ch006

Bhakta, J. N. (2016). Microbial Response against Metal Toxicity. In A. Rathoure & V. Dhatwalia (Eds.), *Toxicity and Waste Management Using Bioremediation* (pp. 75–96). Hershey, PA: IGI Global. doi:10.4018/978-1-4666-9734-8.ch004

Bhatt, P. (2018). Insilico Tools to Study the Bioremediation in Microorganisms. In V. Pathak, & Navneet (Eds.), Handbook of Research on Microbial Tools for Environmental Waste Management (pp. 389-395). Hershey, PA: IGI Global. doi:10.4018/978-1-5225-3540-9.ch018

Bogatu, C., & Leszczynska, D. (2016). Transformation of Nanomaterials in Environment: Surface Activation of SWCNTs during Disinfection of Water with Chlorine. *Journal of Nanotoxicology and Nanomedicine, 1*(1), 45–57. doi:10.4018/JNN.2016010104

Bonmatí-Blasi, A., Cerrillo-Moreno, M., & Riau-Arenas, V. (2017). Systems Based on Physical-Chemical Processes: Nutrient Recovery for Cycle Closure. In Á. Val del Río, J. Campos Gómez, & A. Mosquera Corral (Eds.), *Technologies for the Treatment and Recovery of Nutrients from Industrial Wastewater* (pp. 43–75). Hershey, PA: IGI Global. doi:10.4018/978-1-5225-1037-6.ch003

Cema, G., & Sochacki, A. (2017). Treatment of Landfill Leachate by Anammox Process. In Á. Val del Río, J. Campos Gómez, & A. Mosquera Corral (Eds.), *Technologies for the Treatment and Recovery of Nutrients from Industrial Wastewater* (pp. 290–311). Hershey, PA: IGI Global. doi:10.4018/978-1-5225-1037-6.ch011

Cervantes, F. J., Cuervo-López, F., & Hernández, J. G. (2017). Fundamentals of the Biological Processes for Nitrogen Removal. In Á. Val del Río, J. Campos Gómez, & A. Mosquera Corral (Eds.), *Technologies for the Treatment and Recovery of Nutrients from Industrial Wastewater* (pp. 112–146). Hershey, PA: IGI Global. doi:10.4018/978-1-5225-1037-6.ch005

Chaudhry, S., & Garg, S. (2019). Industrial Wastewater Pollution and Advanced Treatment Techniques. In A. Hussain & S. Ahmed (Eds.), *Advanced Treatment Techniques for Industrial Wastewater* (pp. 74–97). Hershey, PA: IGI Global. doi:10.4018/978-1-5225-5754-8.ch006

Chaurasia, P. K., Bharati, S. L., & Mani, A. (2018). Enzymatic Treatment of Petroleum-Based Hydrocarbons. In V. Pathak, & Navneet (Eds.), Handbook of Research on Microbial Tools for Environmental Waste Management (pp. 396-408). Hershey, PA: IGI Global. doi:10.4018/978-1-5225-3540-9.ch019

Cook, C., & Gude, V. G. (2017). Characteristics of Chitosan Nanoparticles for Water and Wastewater Treatment: Chitosan for Water Treatment. In T. Saleh (Ed.), *Advanced Nanomaterials for Water Engineering, Treatment, and Hydraulics* (pp. 223–261). Hershey, PA: IGI Global. doi:10.4018/978-1-5225-2136-5.ch009

Cuker, B., & Bugyi, G. (2016). The Fundamentals of Water and Natural Waters. In A. McKeown & G. Bugyi (Eds.), *Impact of Water Pollution on Human Health and Environmental Sustainability* (pp. 1–28). Hershey, PA: IGI Global. doi:10.4018/978-1-4666-9559-7.ch001

de la Varga, D., Soto, M., Arias, C. A., van Oirschot, D., Kilian, R., Pascual, A., & Álvarez, J. A. (2017). Constructed Wetlands for Industrial Wastewater Treatment and Removal of Nutrients. In Á. Val del Río, J. Campos Gómez, & A. Mosquera Corral (Eds.), *Technologies for the Treatment and Recovery of Nutrients from Industrial Wastewater* (pp. 202–230). Hershey, PA: IGI Global. doi:10.4018/978-1-5225-1037-6.ch008

Dhatwalia, V. K., & Nanda, M. (2016). Biodegradation of Phenol: Mechanisms and Applications. In A. Rathoure & V. Dhatwalia (Eds.), *Toxicity and Waste Management Using Bioremediation* (pp. 198–214). Hershey, PA: IGI Global. doi:10.4018/978-1-4666-9734-8.ch010

Di Paola, N., Spanò, R., Caldarelli, A., & Vona, R. (2016). Fighting Ecomafias: The Role of Biotech Networks in Achieving Sustainability. In A. Rathoure & V. Dhatwalia (Eds.), *Toxicity and Waste Management Using Bioremediation* (pp. 295–311). Hershey, PA: IGI Global. doi:10.4018/978-1-4666-9734-8.ch015

Dima, J. B., & Zaritzky, N. (2017). Performance of Chitosan Micro/Nanoparticles to Remove Hexavalent Chromium From Residual Water. In T. Saleh (Ed.), *Advanced Nanomaterials for Water Engineering, Treatment, and Hydraulics* (pp. 262–288). Hershey, PA: IGI Global. doi:10.4018/978-1-5225-2136-5.ch010

Dubey, S. K., Hussain, A., & Ajnavi, S. (2019). Environmental Recycling System (ERS): An Emerging Approach to Solid Waste Management. In A. Hussain & S. Ahmed (Eds.), *Advanced Treatment Techniques for Industrial Wastewater* (pp. 225–237). Hershey, PA: IGI Global. doi:10.4018/978-1-5225-5754-8.ch013

Farraji, H. (2018). Phytoremediation of Nitrogen and Phosphorus in Municipal Wastewater by Cyperus alternifolius Planted Constructed Wetland. In V. Pathak, & Navneet (Eds.), Handbook of Research on Microbial Tools for Environmental Waste Management (pp. 146-163). Hershey, PA: IGI Global. doi:10.4018/978-1-5225-3540-9.ch008

Farraji, H., Zaman, N. Q., Vakili, M., & Faraji, H. (2016). Overpopulation and Sustainable Waste Management. *International Journal of Sustainable Economies Management*, *5*(3), 13–36. doi:10.4018/IJSEM.2016070102

Fra-Vázquez, A., & Valenzuela-Heredia, D. (2017). Biogeochemical Cycles of Nitrogen and Phosphorus: Implications of Anthropogenic Activities. In Á. Val del Río, J. Campos Gómez, & A. Mosquera Corral (Eds.), *Technologies for the Treatment and Recovery of Nutrients from Industrial Wastewater* (pp. 1–20). Hershey, PA: IGI Global. doi:10.4018/978-1-5225-1037-6.ch001

Gatt, K. (2016). Youths' Social Traits in Water Management as a Precursor for Good Water Governance. *International Journal of Information Systems and Social Change*, *7*(3), 16–26. doi:10.4018/IJISSC.2016070102

Gupta, C., & Prakash, D. (2016). Novel Bioremediation Methods in Waste Management: Novel Bioremediation Methods. In A. Rathoure & V. Dhatwalia (Eds.), *Toxicity and Waste Management Using Bioremediation* (pp. 141–157). Hershey, PA: IGI Global. doi:10.4018/978-1-4666-9734-8.ch007

Hassan, S. H., & Halim, A. A. (2019). Water Quality Legislation and Regulation. In H. Aziz & S. Abu Amr (Eds.), *Advanced Oxidation Processes (AOPs) in Water and Wastewater Treatment* (pp. 30–45). Hershey, PA: IGI Global. doi:10.4018/978-1-5225-5766-1.ch002

Hilles, A. H., Abu Amr, S. S., Aziz, H. A., & Bashir, M. J. (2019). Advanced Oxidation Processes for Water and Wastewater Treatment: An Introduction. In H. Aziz & S. Abu Amr (Eds.), *Advanced Oxidation Processes (AOPs) in Water and Wastewater Treatment* (pp. 46–69). Hershey, PA: IGI Global. doi:10.4018/978-1-5225-5766-1.ch003

Hivrale, A. U., Pawar, P. K., Rane, N. R., & Govindwar, S. P. (2016). Application of Genomics and Proteomics in Bioremediation. In A. Rathoure & V. Dhatwalia (Eds.), *Toxicity and Waste Management Using Bioremediation* (pp. 97–112). Hershey, PA: IGI Global. doi:10.4018/978-1-4666-9734-8.ch005

Hussain, A., Bhattacharya, A., & Ahmed, A. (2019). Plastic Waste Pollution and Its Management in India: A Review. In A. Hussain & S. Ahmed (Eds.), *Advanced Treatment Techniques for Industrial Wastewater* (pp. 62–73). Hershey, PA: IGI Global. doi:10.4018/978-1-5225-5754-8.ch005

Ibrahim, N., Zainal, S. F., & Aziz, H. A. (2019). Application of UV-Based Advanced Oxidation Processes in Water and Wastewater Treatment. In H. Aziz & S. Abu Amr (Eds.), *Advanced Oxidation Processes (AOPs) in Water and Wastewater Treatment* (pp. 384–414). Hershey, PA: IGI Global. doi:10.4018/978-1-5225-5766-1.ch014

Jagdale, S., & Chabukswar, A. (2016). Phyto-Remediation: Using Plants to Clean Up Soils: Phyto-Remediation. In A. Rathoure & V. Dhatwalia (Eds.), *Toxicity and Waste Management Using Bioremediation* (pp. 215–235). Hershey, PA: IGI Global. doi:10.4018/978-1-4666-9734-8.ch011

Kamaruddin, M. A. (2019). Advanced Oxidation Processes (AOPs) in Landfill Leachate Treatment. In H. Aziz & S. Abu Amr (Eds.), *Advanced Oxidation Processes (AOPs) in Water and Wastewater Treatment* (pp. 355–383). Hershey, PA: IGI Global. doi:10.4018/978-1-5225-5766-1.ch013

Kaur, A. (2017). Applications of Nanomaterials for Water Treatment: A Future Avenue. In T. Saleh (Ed.), *Advanced Nanomaterials for Water Engineering, Treatment, and Hydraulics* (pp. 289–304). Hershey, PA: IGI Global. doi:10.4018/978-1-5225-2136-5.ch011

Kehinde, F. O., Abu Amr, S. S., & Aziz, H. A. (2019). Application of Persulfate in Textile Wastewater Treatment. In H. Aziz & S. Abu Amr (Eds.), *Advanced Oxidation Processes (AOPs) in Water and Wastewater Treatment* (pp. 70–98). Hershey, PA: IGI Global. doi:10.4018/978-1-5225-5766-1.ch004

Khelfi, A. (2018). Sources of Air Pollution. In V. Pathak, & Navneet (Eds.), Handbook of Research on Microbial Tools for Environmental Waste Management(pp. 220-258). Hershey, PA: IGI Global. doi:10.4018/978-1-5225-3540-9.ch011

Khelfi, A. (2018). Sources of Soil Pollution. In V. Pathak, & Navneet (Eds.), Handbook of Research on Microbial Tools for Environmental Waste Management(pp. 190-218). Hershey, PA: IGI Global. doi:10.4018/978-1-5225-3540-9.ch010

Khelfi, A. (2019). Sources of Groundwater Pollution. In A. Hussain & S. Ahmed (Eds.), *Advanced Treatment Techniques for Industrial Wastewater* (pp. 177–210). Hershey, PA: IGI Global. doi:10.4018/978-1-5225-5754-8.ch011

Koli, S. K., & Hussain, A. (2019). Status of Electronic Waste Management in India: A Review. In A. Hussain & S. Ahmed (Eds.), *Advanced Treatment Techniques for Industrial Wastewater* (pp. 238–250). Hershey, PA: IGI Global. doi:10.4018/978-1-5225-5754-8.ch014

Kulkarni, A. G., & Vedamurthy, A. B. (2018). Cellulolytic Microorganisms: A Review. In V. Pathak, & Navneet (Eds.), Handbook of Research on Microbial Tools for Environmental Waste Management (pp. 34-47). Hershey, PA: IGI Global. doi:10.4018/978-1-5225-3540-9.ch002

Kumar, P. S., & C., F. C. (2019). Soil Bioremediation Techniques. In A. Hussain, & S. Ahmed (Eds.), *Advanced Treatment Techniques for Industrial Wastewater* (pp. 35-50). Hershey, PA: IGI Global. doi:10.4018/978-1-5225-5754-8.ch003

Kumar, P. S., & Saravanan, A. (2019). Water Pollutants and Their Removal Techniques. In A. Hussain & S. Ahmed (Eds.), *Advanced Treatment Techniques for Industrial Wastewater* (pp. 114–133). Hershey, PA: IGI Global. doi:10.4018/978-1-5225-5754-8.ch008

Kumar, P. S., & Yaashikaa, P. R. (2019). Agriculture Pollution. In A. Hussain & S. Ahmed (Eds.), *Advanced Treatment Techniques for Industrial Wastewater* (pp. 134–154). Hershey, PA: IGI Global. doi:10.4018/978-1-5225-5754-8.ch009

Laha, S. D., Dutta, K., & Kundu, P. P. (2018). Biodegradation of Low Density Polyethylene Films. In V. Pathak, & Navneet (Eds.), Handbook of Research on Microbial Tools for Environmental Waste Management (pp. 282-318). Hershey, PA: IGI Global. doi:10.4018/978-1-5225-3540-9.ch014

Manyuchi, M. M., Muzenda, E., & Mbohwa, C. (2018). Design and Development of a Sanitary Landfill for Low Income Countries for Optimal Waste Management. In V. Pathak, & Navneet (Eds.), Handbook of Research on Microbial Tools for Environmental Waste Management (pp. 373-388). Hershey, PA: IGI Global. doi:10.4018/978-1-5225-3540-9.ch017

Mbuh, M. J., Houser, P. R., & Heidari, A. (2016). Water Quality Estimation Using Combined Water Chemistry and Field Spectroscopy in the Shenandoah River, Virginia. *International Journal of Applied Geospatial Research*, *7*(2), 14–37. doi:10.4018/ijagr.2016040102

McKeown, A. E. (2016). Nurses, Healthcare, and Environmental Pollution and Solutions: Breaking the Cycle of Harm. In A. McKeown & G. Bugyi (Eds.), *Impact of Water Pollution on Human Health and Environmental Sustainability* (pp. 174–197). Hershey, PA: IGI Global. doi:10.4018/978-1-4666-9559-7.ch008

Mobin, R., Chisti, H., Rangreez, T. A., Bashir, R., & Najar, A. A. (2019). Pesticidal Pollution. In A. Hussain & S. Ahmed (Eds.), *Advanced Treatment Techniques for Industrial Wastewater* (pp. 155–176). Hershey, PA: IGI Global. doi:10.4018/978-1-5225-5754-8.ch010

Mohamed, Z. (2016). Cyanobacterial Toxins in Water Sources and Their Impacts on Human Health. In A. McKeown & G. Bugyi (Eds.), *Impact of Water Pollution on Human Health and Environmental Sustainability* (pp. 120–149). Hershey, PA: IGI Global. doi:10.4018/978-1-4666-9559-7.ch006

Mojiri, A., Ziyang, L., Hui, W., & Gholami, A. (2019). Concentrated Landfill Leachate Treatment by Electro-Ozonation. In H. Aziz & S. Abu Amr (Eds.), *Advanced Oxidation Processes (AOPs) in Water and Wastewater Treatment* (pp. 150–170). Hershey, PA: IGI Global. doi:10.4018/978-1-5225-5766-1.ch007

Mujere, N. (2016). Water Hyacinth: Characteristics, Problems, Control Options, and Beneficial Uses. In A. McKeown & G. Bugyi (Eds.), *Impact of Water Pollution on Human Health and Environmental Sustainability* (pp. 343–361). Hershey, PA: IGI Global. doi:10.4018/978-1-4666-9559-7.ch015

Mujere, N., & Isidro, M. (2016). Impacts of Artisanal and Small-Scale Gold Mining on Water Quality in Mozambique and Zimbabwe. In A. McKeown & G. Bugyi (Eds.), *Impact of Water Pollution on Human Health and Environmental Sustainability* (pp. 101–119). Hershey, PA: IGI Global. doi:10.4018/978-1-4666-9559-7.ch005

Nazir, R. (2017). Advanced Nanomaterials for Water Engineering and Treatment: Nano-Metal Oxides and Their Nanocomposites. In T. Saleh (Ed.), *Advanced Nanomaterials for Water Engineering, Treatment, and Hydraulics* (pp. 84–126). Hershey, PA: IGI Global. doi:10.4018/978-1-5225-2136-5.ch005

Obeng, P. A., Oduro-Kwarteng, S., Keraita, B., Bregnhøj, H., Abaidoo, R. C., Awuah, E., & Konradsen, F. (2016). On-Site Sanitation Systems for Low-Income Countries: Technical Guidelines for Groundwater Pollution and Nuisance Control. In A. McKeown & G. Bugyi (Eds.), *Impact of Water Pollution on Human Health and Environmental Sustainability* (pp. 319–342). Hershey, PA: IGI Global. doi:10.4018/978-1-4666-9559-7.ch014

Olajubu, E. A., Ajayi, G., Oke, I., & Asahiah, F. O. (2017). An ANN Model for Predicting the Quantity of Lead and Cadmium Ions in Industrial Wastewater. *International Journal of Information Communication Technologies and Human Development*, *9*(4), 32–44. doi:10.4018/IJICTHD.2017100103

Paris, R. S. (2016). Issues with Water Quality: How Do We Get Our Fellow Citizens to Care? In A. McKeown & G. Bugyi (Eds.), *Impact of Water Pollution on Human Health and Environmental Sustainability* (pp. 199–214). Hershey, PA: IGI Global. doi:10.4018/978-1-4666-9559-7.ch009

Pedemonte, D. C., Frison, N., Tayà, C., Ponsa, S., & Fatone, F. (2017). Chemical and Biological Processes for Nutrients Removal and Recovery. In Á. Val del Río, J. Campos Gómez, & A. Mosquera Corral (Eds.), *Technologies for the Treatment and Recovery of Nutrients from Industrial Wastewater* (pp. 76–111). Hershey, PA: IGI Global. doi:10.4018/978-1-5225-1037-6.ch004

Prabu, S. L., Suriyaprakash, T., Kandasamy, R., & Rathinasabapathy, T. (2016). Effective Waste Water Treatment and its Management. In A. Rathoure & V. Dhatwalia (Eds.), *Toxicity and Waste Management Using Bioremediation* (pp. 312–334). Hershey, PA: IGI Global. doi:10.4018/978-1-4666-9734-8.ch016

Priyanka, M., Kumar, D., Shankar, U., Yadav, A., & Yadav, K. (2018). Agricultural Waste Management for Bioethanol Production. In V. Pathak, & Navneet (Eds.), Handbook of Research on Microbial Tools for Environmental Waste Management(pp. 1-33). Hershey, PA: IGI Global. doi:10.4018/978-1-5225-3540-9.ch001

Rajasekaran, R. N., Puddoo, H., & Nguyenhuynh, T. N. (2019). An Overview of Treatment of Antibiotics Using Advanced Oxidation Process. In H. Aziz & S. Abu Amr (Eds.), *Advanced Oxidation Processes (AOPs) in Water and Wastewater Treatment* (pp. 226–260). Hershey, PA: IGI Global. doi:10.4018/978-1-5225-5766-1.ch010

Rangreez, T. A., Mobin, R., Chisti, H., Bashir, R., & Ara, T. (2019). Advanced Water Treatment Systems and Their Applications. In A. Hussain & S. Ahmed (Eds.), *Advanced Treatment Techniques for Industrial Wastewater* (pp. 1–18). Hershey, PA: IGI Global. doi:10.4018/978-1-5225-5754-8.ch001

Rastogi, M., & Nandal, M. (2018). Composting an Emerging Technology for Solid Waste Management in India. In V. Pathak, & Navneet (Eds.), Handbook of Research on Microbial Tools for Environmental Waste Management (pp. 105-126). Hershey, PA: IGI Global. doi:10.4018/978-1-5225-3540-9.ch006

Rathoure, A. K. (2016). Heavy Metal Pollution and its Management: Bioremediation of Heavy Metal. In A. Rathoure & V. Dhatwalia (Eds.), *Toxicity and Waste Management Using Bioremediation* (pp. 27–50). Hershey, PA: IGI Global. doi:10.4018/978-1-4666-9734-8.ch002

Sadegh, H., & Ali, G. A. (2019). Potential Applications of Nanomaterials in Wastewater Treatment: Nanoadsorbents Performance. In A. Hussain & S. Ahmed (Eds.), *Advanced Treatment Techniques for Industrial Wastewater* (pp. 51–61). Hershey, PA: IGI Global. doi:10.4018/978-1-5225-5754-8.ch004

Said, S., Hussain, A., & Sharma, G. (2018). Water Quality Mapping of Yamuna River Stretch Passing Through Delhi State Using High Resolution Geoeye-2 Imagery. *International Journal of Applied Geospatial Research, 9*(4), 23–35. doi:10.4018/IJAGR.2018100102

Saleh, T. A. (2017). An Overview of Nanomaterials for Water Technology. In T. Saleh (Ed.), *Advanced Nanomaterials for Water Engineering, Treatment, and Hydraulics* (pp. 1–12). Hershey, PA: IGI Global. doi:10.4018/978-1-5225-2136-5.ch001

Saleh, T. A., Adio, S. O., Parthasarathy, P., & Danmaliki, G. I. (2017). Scientific Insights Into Modified and Non-Modified Biomaterials for Sorption of Heavy Metals From Water. In T. Saleh (Ed.), *Advanced Nanomaterials for Water Engineering, Treatment, and Hydraulics* (pp. 13–39). Hershey, PA: IGI Global. doi:10.4018/978-1-5225-2136-5.ch002

Saleh, T. A., & Danmaliki, G. I. (2017). Polymer Consumption, Environmental Concerns, Possible Disposal Options, and Recycling for Water Treatment. In T. Saleh (Ed.), *Advanced Nanomaterials for Water Engineering, Treatment, and Hydraulics* (pp. 200–222). Hershey, PA: IGI Global. doi:10.4018/978-1-5225-2136-5.ch008

Saleh, T. A., Majeed, S., Nayak, A., & Bhushan, B. (2017). Principles and Advantages of Microwave-Assisted Methods for the Synthesis of Nanomaterials for Water Purification. In T. Saleh (Ed.), *Advanced Nanomaterials for Water Engineering, Treatment, and Hydraulics* (pp. 40–57). Hershey, PA: IGI Global. doi:10.4018/978-1-5225-2136-5.ch003

Satpathy, R. (2018). Microbial Bioreactor Systems for Dehalogenation of Organic Pollutants. In V. Pathak, & Navneet (Eds.), Handbook of Research on Microbial Tools for Environmental Waste Management (pp. 319-340). Hershey, PA: IGI Global. doi:10.4018/978-1-5225-3540-9.ch015

Shankara, S., Dupadahalli, K., Vijayakumar, M. H., & Gaddad, S. M. (2016). Decolorization of Direct Blue: 14 Dye by Thermoalkalophilic Aerobic Bacillus sp. In A. Rathoure & V. Dhatwalia (Eds.), *Toxicity and Waste Management Using Bioremediation* (pp. 279–294). Hershey, PA: IGI Global. doi:10.4018/978-1-4666-9734-8.ch014

Sharma, A., Kaur, M., Katnoria, J. K., & Nagpal, A. K. (2016). Heavy Metal Pollution: A Global Pollutant of Rising Concern. In A. Rathoure & V. Dhatwalia (Eds.), *Toxicity and Waste Management Using Bioremediation* (pp. 1–26). Hershey, PA: IGI Global. doi:10.4018/978-1-4666-9734-8.ch001

Shikuku, V. O., & Achieng', G. O. (2019). Occurrence and Fate of Selected Heavy Metals in a Conventional Municipal Wastewater Treatment Plant in Kisumu City, Kenya: A Case Study. In A. Hussain & S. Ahmed (Eds.), *Advanced Treatment Techniques for Industrial Wastewater* (pp. 211–224). Hershey, PA: IGI Global. doi:10.4018/978-1-5225-5754-8.ch012

Shikuku, V. O., Kowenje, C. O., & Nyairo, W. N. (2017). Fundamentals and Sources of Magnetic Nanocomposites and Their Sorption Properties. In T. Saleh (Ed.), *Advanced Nanomaterials for Water Engineering, Treatment, and Hydraulics* (pp. 58–83). Hershey, PA: IGI Global. doi:10.4018/978-1-5225-2136-5.ch004

Shikuku, V. O., Nyairo, W. N., & Kowenje, C. O. (2019). Preparation and Application of Biochars for Organic and Microbial Control in Wastewater Treatment Regimes. In A. Hussain & S. Ahmed (Eds.), *Advanced Treatment Techniques for Industrial Wastewater* (pp. 19–34). Hershey, PA: IGI Global. doi:10.4018/978-1-5225-5754-8.ch002

Shrivas, K., Ghosale, A., & Maji, P. (2017). Advanced Nanomaterials for the Removal of Chemical Substances and Microbes From Contaminated and Waste Water. In T. Saleh (Ed.), *Advanced Nanomaterials for Water Engineering, Treatment, and Hydraulics* (pp. 127–161). Hershey, PA: IGI Global. doi:10.4018/978-1-5225-2136-5.ch006

Simmonds, J., Gómez, J. A., & Ledezma, A. (2018). Statistical and Data Mining Techniques for Understanding Water Quality Profiles in a Mining-Affected River Basin. *International Journal of Agricultural and Environmental Information Systems*, *9*(2), 1–19. doi:10.4018/IJAEIS.2018040101

Singh, R., & Chauhan, M. (2016). Effective Management of Agro-Industrial Residues as Composting in Mushroom Industry and Utilization of Spent Mushroom Substrate for Bioremediation. In A. Rathoure & V. Dhatwalia (Eds.), *Toxicity and Waste Management Using Bioremediation* (pp. 158–177). Hershey, PA: IGI Global. doi:10.4018/978-1-4666-9734-8.ch008

Smith, M. C. (2016). Design of the Education Campaign: Creating a Motivated, Environmentally Concerned Public through Improved Scientific Literacy. In A. McKeown & G. Bugyi (Eds.), *Impact of Water Pollution on Human Health and Environmental Sustainability* (pp. 215–239). Hershey, PA: IGI Global. doi:10.4018/978-1-4666-9559-7.ch010

Starkl, M., Brunner, N., Hauser, A. W., Feil, M., & Kasan, H. (2018). Addressing Sustainability of Sanitation Systems: Can it be Standardized? *International Journal of Standardization Research*, *16*(1), 39–51. doi:10.4018/IJSR.2018010103

Stefanakis, A. I. (2016). Constructed Wetlands: Description and Benefits of an Eco-Tech Water Treatment System. In A. McKeown & G. Bugyi (Eds.), *Impact of Water Pollution on Human Health and Environmental Sustainability* (pp. 281–303). Hershey, PA: IGI Global. doi:10.4018/978-1-4666-9559-7.ch012

Stefanakis, A. I., & Becker, J. A. (2016). A Review of Emerging Contaminants in Water: Classification, Sources, and Potential Risks. In A. McKeown & G. Bugyi (Eds.), *Impact of Water Pollution on Human Health and Environmental Sustainability* (pp. 55–80). Hershey, PA: IGI Global. doi:10.4018/978-1-4666-9559-7.ch003

Stratigea, A., Leka, A., & Panagiotopoulou, M. (2017). In Search of Indicators for Assessing Smart and Sustainable Cities and Communities' Performance. *International Journal of E-Planning Research, 6*(1), 43–73. doi:10.4018/IJEPR.2017010103

Tajarudin, H. A., Makhtar, M. M., Azmi, M. S., Zainuddin, N. I., & Ali, D. H. (2019). Introduction to Water and Wastewater Treatment. In H. Aziz & S. Abu Amr (Eds.), *Advanced Oxidation Processes (AOPs) in Water and Wastewater Treatment* (pp. 1–29). Hershey, PA: IGI Global. doi:10.4018/978-1-5225-5766-1.ch001

Tischler, D., Qi, J., Ngo, A. C., & Schlömann, M. (2018). Microbial Degradation of Azo Dyes: The Role of Azoreductase to Initiate Degradation. In V. Pathak, & Navneet (Eds.), Handbook of Research on Microbial Tools for Environmental Waste Management (pp. 341-371). Hershey, PA: IGI Global. doi:10.4018/978-1-5225-3540-9.ch016

Turcea, V. C., & Mihai, M. M. (2019). The Impact of CO2 Emissions From Waste on Human Health: A Consequence of Globalization. *International Journal of Sustainable Economies Management, 8*(1), 17–28. doi:10.4018/IJSEM.2019010102

Umesh, M., & Basheer, T. (2018). Microbe Mediated Bioconversion of Fruit Waste Into Value Added Products: Microbes in Fruit Waste Management. In V. Pathak, & Navneet (Eds.), Handbook of Research on Microbial Tools for Environmental Waste Management (pp. 57-78). Hershey, PA: IGI Global. doi:10.4018/978-1-5225-3540-9.ch004

Val del Río, Á., Fernández, P. C., Gómez, J. L., & Mosquera-Corral, A. (2017). Nutrients Pollution in Water Bodies: Related Legislation in Europe and the United States. In Á. Val del Río, J. Campos Gómez, & A. Mosquera Corral (Eds.), *Technologies for the Treatment and Recovery of Nutrients from Industrial Wastewater* (pp. 21–42). Hershey, PA: IGI Global. doi:10.4018/978-1-5225-1037-6.ch002

Val del Río, Á., Fuentes, A. P., Giustinianovich, E. A., Gomez, J. L., & Mosquera-Corral, A. (2017). Anammox Process: Technologies and Application to Industrial Effluents. In Á. Val del Río, J. Campos Gómez, & A. Mosquera Corral (Eds.), *Technologies for the Treatment and Recovery of Nutrients from Industrial Wastewater* (pp. 264–289). Hershey, PA: IGI Global. doi:10.4018/978-1-5225-1037-6.ch010

Valverde-Pérez, B., Flores-Alsina, X., Vangsgaard, A. K., & Mauricio-Iglesias, M. (2017). Modelling and Control of Nitrogen and Phosphorus Removing Systems. In Á. Val del Río, J. Campos Gómez, & A. Mosquera Corral (Eds.), *Technologies for the Treatment and Recovery of Nutrients from Industrial Wastewater* (pp. 174–201). Hershey, PA: IGI Global. doi:10.4018/978-1-5225-1037-6.ch007

Verma, A., & Shukla, P. K. (2016). A Prospective Study on Emerging Role of Phytoremediation by Endophytic Microorganisms. In A. Rathoure & V. Dhatwalia (Eds.), *Toxicity and Waste Management Using Bioremediation* (pp. 236–265). Hershey, PA: IGI Global. doi:10.4018/978-1-4666-9734-8.ch012

Vipond, E., & Rahman, P. K. (2018). Anaerobic Digestion Enhancement With Microbial Electrolysis Cells: Is Biomethane Production the Direction to Go for Commercialization? In V. Pathak, & Navneet (Eds.), Handbook of Research on Microbial Tools for Environmental Waste Management (pp. 259-269). Hershey, PA: IGI Global. doi:10.4018/978-1-5225-3540-9.ch012

Voulvoulis, N., & Georges, K. (2016). Industrial and Agricultural Sources and Pathways of Aquatic Pollution. In A. McKeown & G. Bugyi (Eds.), *Impact of Water Pollution on Human Health and Environmental Sustainability* (pp. 29–54). Hershey, PA: IGI Global. doi:10.4018/978-1-4666-9559-7.ch002

Wahla, V., & Shukla, S. (2018). Role of Microorganisms in Bioremediation of Pesticides. In V. Pathak, & Navneet (Eds.), Handbook of Research on Microbial Tools for Environmental Waste Management (pp. 164-189). Hershey, PA: IGI Global. doi:10.4018/978-1-5225-3540-9.ch009

Willems, R. C., & MacDonald, S. A. (2016). Access to Safe Drinking Water: A Low Income Sustainable Water Purification System. In A. McKeown & G. Bugyi (Eds.), *Impact of Water Pollution on Human Health and Environmental Sustainability* (pp. 304–318). Hershey, PA: IGI Global. doi:10.4018/978-1-4666-9559-7.ch013

Zakaria, S. N. (2019). Ozonation With Catalyst in Landfill Leachate Treatment. In H. Aziz & S. Abu Amr (Eds.), *Advanced Oxidation Processes (AOPs) in Water and Wastewater Treatment* (pp. 324–354). Hershey, PA: IGI Global. doi:10.4018/978-1-5225-5766-1.ch012

Zaman, A., Das, P., & Banerjee, P. (2016). Biosorption of Dye Molecules. In A. Rathoure & V. Dhatwalia (Eds.), *Toxicity and Waste Management Using Bioremediation* (pp. 51–74). Hershey, PA: IGI Global. doi:10.4018/978-1-4666-9734-8.ch003

Ziemba, R., Cuker, B. E., Stein, J., Meuninck, R., & Wan, J. A. (2016). Health Professionals Can Protect Water Quality: Tools for Educators, Advocates, and Practitioners. In A. McKeown & G. Bugyi (Eds.), *Impact of Water Pollution on Human Health and Environmental Sustainability* (pp. 240–280). Hershey, PA: IGI Global. doi:10.4018/978-1-4666-9559-7.ch011

About the Author

Yazan Khalid Abed-Allah Migdadi is an Associate Professor of Operations Management at Management and Marketing Department, Qatar University, Qatar. He was awarded Ph.D. in Operations Management from Bradford University, UK. Also, awarded B.A. and M.B.A. from Yarmouk University, Jordan. He worked as an assistant professor and associate professor of operations management at Yarmouk University, Jordan. Also, he worked as a teaching assistant and a researcher of operations and information management at Bradford University, School of Management, UK. His main research interest is reporting operations strategy: best practices, taxonomies, process and typologies. Also, reporting green operations strategy: taxonomies and best practices. Dr. Yazan published more than 30 publications. The publications appeared in leading databases, as IGI Global, Emerald, Taylor & Francis, Wiley & Sons, and Inderscience. Dr. Yazan concerns about reporting the practices of many sectors, as aviation, information and communication technology, banking and healthcare.

Index

A

AC equipment 65, 81
Air Transport 1, 18, 20, 43, 88, 91, 105-106, 184
air transport industry 88, 91
aircraft parts 91, 106
alternative fuels 7, 21
automated energy-saving 65, 81
automation system 78, 82
auxiliary power 37, 39-40, 45
aviation sector 2, 20, 46, 112, 172-173

C

CAD procedures 143, 149-150
civil aviation authority 2, 131-133, 136, 143, 148-150, 184
classification scheme 5, 7, 20-21, 155, 176
clean paper 6, 89, 105
clustering analysis 5, 154-157, 160-161, 163, 165
cooling towers 65, 122, 126

D

day-to-day operations 82
direct GHG emissions 1, 5, 7, 17-19, 21-22, 24, 27-35, 37-43, 45-47, 52-54
diverse actions 43, 47

E

Eco Shine 126
effective green operations 1, 9, 18, 29, 61, 88, 111, 153-154, 159, 161, 165, 172-173, 175-176, 178, 180-185, 190-192, 205-206, 212
electrical carts 41, 46
electrical taxiing 149-150
electricity meters 76, 80-81
energy saving 8, 60, 64-67, 74-76, 79-81, 83, 158, 175
environmental effect 172-173

F

flight operations 7-8, 17-18, 21-24, 29-30, 32-33, 35, 37-38, 42-43, 87, 146, 199
flight rescheduling 141, 143, 149-150
fuel management 21, 24, 35, 37, 44-47, 158-159

G

GHG emissions 1, 3, 5-8, 17-22, 24, 27-35, 37-43, 45-47, 52-54, 62, 64, 81, 153, 158, 161, 164-165, 175, 178, 180-181, 184-185, 206
GHG2 emissions 6, 8, 60-64, 66-74, 76-79, 82-83, 157-158, 200-201
green actions 7-8, 18, 61, 150, 186
green operations 1, 4, 7, 9, 18-19, 29, 43, 47, 61-63, 83, 88-89, 106, 111, 127, 132-133, 136, 149-150, 153-159, 161, 165, 169, 172-176, 178, 180-185, 190-192, 198-202, 205-206, 212
green operations strategy 1, 4, 9, 43, 47, 83, 106, 153-154, 156-159, 161, 165, 172-176, 178, 180-183, 185, 190-192,

199, 202, 205-206, 212
ground operations 7, 17, 21-24, 29-30, 40-41, 43, 46-47, 114
ground support 5, 17-18, 20, 24, 62, 64

H

heat pumps 81
Heat, Ventilation, Air Conditioning (HVAC) 81
heating/cooling modifications 65, 81

I

indirect GHG2 emissions 8, 60-64, 66-74, 76-79, 82-83, 157

K

K-means clustering 5, 154-155, 157, 160-161, 163, 165

L

land-use planning 3, 8-9, 132, 134-136, 149
LEED standards 65-66, 76-77, 80, 82-83
lighting systems 63, 81
low-to-moderate emphasizer 164, 175, 184

M

metric tons 6, 20, 24, 62, 67

N

night flight rescheduling 143, 149-150
noise management 7-9, 131-136, 138, 140, 149, 153, 161, 165, 176, 178, 180-186, 205-207
noise management strategies 131, 140, 149

O

on-off function 65, 81
operations strategy 1, 4, 9, 43, 46-47, 83, 106, 146, 153-159, 161, 165, 172-176, 178, 180-183, 185, 190-192, 198-200, 202, 205-206, 212

S

strategic patterns 8-9, 17-19, 22, 27, 29, 32-35, 37-38, 40-42, 53, 60-61, 63, 66, 69, 71, 73-74, 76-78, 82, 88, 92, 95, 97, 99-101, 104-105, 110-113, 116, 118, 120-122, 124, 131-132, 136-138, 140-141, 144, 148-150, 159, 169, 173, 176, 178, 191
strategy patterns 1, 4-5, 8-9, 18, 21, 23, 29, 32, 40, 45-47, 61-62, 64, 66, 73, 76, 83, 87, 89-92, 97, 99, 105-107, 111-113, 118, 120, 124, 127, 133-134, 137, 140, 150, 153-155, 157-161, 163-165, 172-178, 180-187, 190-192, 194, 197-201, 205-208, 212
sustainable energy 8, 60, 64-66, 68, 74-76, 78-79, 81-83, 158, 200
Sustainable Transportation 2, 4, 18, 61

W

wastes management 7, 87, 90, 93, 95, 97, 99-101, 104, 106-107, 201
wastes reduction 8, 87-88, 91, 97
water consumption 1, 5-6, 110-114, 118-121, 124, 127, 157, 175, 202
water management 7, 113-114, 120, 122, 153, 161, 164-165, 175-176, 178, 180, 182, 185, 202, 205, 207
water saving 2, 8-9, 110-113, 116, 118-122, 124-127
water saving strategies 118, 127
world traffic 2, 173, 184

www.ingramcontent.com/pod-product-compliance
Lightning Source LLC
LaVergne TN
LVHW081323300926
849161LV00008B/1860